About Island Press

Since 1984, the nonprofit organization Island Press has been stimulating, shaping, and communicating ideas that are essential for solving environmental problems worldwide. With more than 1,000 titles in print and some 30 new releases each year, we are the nation's leading publisher on environmental issues. We identify innovative thinkers and emerging trends in the environmental field. We work with world-renowned experts and authors to develop cross-disciplinary solutions to environmental challenges.

Island Press designs and executes educational campaigns, in conjunction with our authors, to communicate their critical messages in print, in person, and online using the latest technologies, innovative programs, and the media. Our goal is to reach targeted audiences—scientists, policy makers, environmental advocates, urban planners, the media, and concerned citizens—with information that can be used to create the framework for long-term ecological health and human well-being.

Island Press gratefully acknowledges major support from The Bobolink Foundation, Caldera Foundation, The Curtis and Edith Munson Foundation, The Forrest C. and Frances H. Lattner Foundation, The JPB Foundation, The Kresge Foundation, The Summit Charitable Foundation, Inc., and many other generous organizations and individuals.

The opinions expressed in this book are those of the author(s) and do not necessarily reflect the views of our supporters.

MAKING CLIMATE TECH WORK

MAKING CLIMATE TECH WORK

Policies that Drive Innovation

Alon Tal

 ISLANDPRESS | Washington | Covelo

Library of Congress Control Number: 2023948842

All Island Press books are printed on environmentally responsible materials.

Manufactured in the United States of America
10 9 8 7 6 5 4 3 2 1

Keywords: Carbon contracts; Carbon taxes; Conference of the Parties (COP); Decarbonization; Electric vehicles; Emission trading systems; Emissions reduction; Energy efficiency; Feed-in tariff; Inflation Reduction Act (IRA); Innovation; Just transition; Mitigation; Net zero; Net-metering; Nudges; Public policy; Public procurement; Renewable energy; Research and development; Social cost of carbon; Solar power; Sustainable development; United Nations Framework Convention on Climate Change; Wind energy

This is book is dedicated to the extraordinary
Thomas Stanton, Yoni Dabas, and Itai Gat.

☙❧

We are so lucky to have you on the team.

CONTENTS

PREFACE

The origins of this book can be traced back to several events. Quite unexpect-
edly, during the lockdown days of 2020, I found myself cc'd in a forty-person
forum with vaunted academics from universities around the world. These were
serious people who think seriously about sustainability. How I was included in
such a storied group of scholars remains a mystery. Nonetheless, I was happy to
be a fly on the computer screens of such auspicious experts.

The conversations meandered among many of the pressing ecological issues
of our time. A common concern was the perils of climate change. As the dialogue
drifted in the direction of platitudes regarding the importance of increasing
global awareness about the severity of global warming, Bill Rees weighed in.

For those still working on their environmental literacy, University of British Columbia researcher William Rees is something of an ecological superstar.
Perhaps chief among many professional distinctions is his research with a gifted
graduate student, Mathis Wackernagel, to develop the theory and methods for
measuring ecological footprints. Their approach continues to provide an accessible and rigorous measure for characterizing the Earth's carrying capacity.

Professor Rees's candor immediately brought the conversation down to
earth. He observed that it was highly unlikely that the climate crisis would be
solved by educational and consciousness-raising initiatives. To paraphrase his
point, Rees explained to the participants that there probably was no group of
individuals on the planet who better understood the adverse effects of global
warming on the planet than the present company. But then he asked rhetorically,
"How many of you have ever voluntarily passed up an international flight for an
important visit because of the associated emissions?"

He clarified that he was not trying to be sanctimonious. Indeed, Rees
acknowledged that he, too, is given to rationalizing periodic carbon-intensive
individual trips for some "greater good." His point was that if such an illustrious
assemblage of climate mavens and sustainability authorities are disinclined to
make meaningful sacrifices in their own mobility to save the planet, then what
could realistically be expected from the general public?

The argument resonated strongly with me. Like any well-meaning environmentalist, I rarely experience a day that is not punctuated by various acts of ecological inconsistency or even hypocrisy. I, too, have long harbored a nagging realization that if humanity does not figure out how to fuel commercial airlines with hydrogen or some very low-carbon alternative, visions of a carbon-neutral emissions equilibrium will remain elusive. I was also well aware that the innovation required to achieve a rapid transformation in aviation fuel would not happen by itself. It will only emerge in response to a clear set of market signals and policy prescriptions that keep scientists, engineers, and airlines engaged.

For a long time I have seen the gap grow larger between where the world needs to be in reducing our collective emissions and where we actually are. The old fossil fuel economy tarries, even though, with every passing day, the risk of nonlinear climate cacophony increases. There are many reasons for this. When I served as chair of the Environment, Climate, and Health subcommittee of the Knesset, Israel's parliament, it was easy to see them up close, and the motivation for this book began to fully coalesce.

To start with, disingenuous disinformation campaigns, financed by the "merchants of doubt" from the fossil fuel industry, have for decades sought to undermine the legitimacy of climate science. Tragically, they have been highly effective. Those interests have not disappeared and continue to pose political challenges to decarbonization progress and future sustainability around the world. In addition to the usual bureaucratic obstacles associated with developing expensive, disruptive, and even radical technologies, climate tech companies must also overcome extremely powerful old-economy adversaries who are not interested in becoming obsolete and irrelevant in the new low-carbon world. For instance, the auto industry in Germany is often blamed for the government's failure to enact effective climate policies for the transport sector, while the fossil fuel industry is accused of sabotaging Japanese support for key decarbonizing targets.

Less insidious, but no less formidable, are obstacles created by well-meaning government officials, committed to formal procedures and safeguards, who do not always understand the importance of flexibility and speedy decision-making when approving critical decarbonizing technology. This a common phenomenon: the US Environmental Protection Agency faces criticism for its sluggishness in approving carbon-capture injection wells. There are companies willing to pay a green premium for low-carbon cement but face inflexible regulations around the chemistry of cement composition. Novel ingredients used in plant-based or cultured meat products can trigger the red flag of food safety regulations, leading to massive delays. Researchers seeking a permit to test a new renewable energy application or energy storage project must navigate the many hurdles associated with land-use planning. In short, legitimate safeguards designed to protect the environment cannot distinguish between projects whose delay is detrimental to environmental interests and those that need to be streamlined.

As chair of the Climate Subcommittee in the Israeli parliament, I visited many promising climate tech companies around the country, from the cultured meat and milk research laboratories of Aleph Farms and Remilk to experimental agrovoltaic systems in grape vineyards and UBQ's "alchemy-like" pilot plant, which creates thermoplastics from unsorted household waste that would otherwise be sent to landfills. These companies tend to be the lucky ones; while domestic incentives and local climate tech policies may be inadequate to give them the economies of scale they need at home in Israel, they will ultimately find a place in the European climate ecosystems. There, I believe they will flourish and emerge as important decarbonization endeavors.

By way of contrast, there are hundreds—possibly thousands—of equally promising, yet less fortunate climate tech startups around the world. Many have extraordinary ideas but will end up fading away or going bankrupt before their unique technologies can make a contribution to the collective global mitigation effort. Better public policies would give them a much better chance to succeed.

In recent years, there have been signs that things are starting to change and move in a sustainable direction. The correct policies, however, could make things happen much, much faster. And the world desperately needs to move faster.

Failure to expedite promising climate tech opportunities is not simply a case of overzealous planning commissions' protecting agricultural fields from turbines and solar farms. Rather, it reflects a general lack of urgency in many governments' perceptions about the planet's climate future—the kind of urgency that compels governments to produce game-changing military technologies in wartime.

It is this very lack of urgency, this complacency, that leads to a shortage of funding for research and development at critical junctures in a startup company's development. What policy support there is might appear as anemic government tax credits, interminable permitting processes, or unsustainable procurement preferences, all of which make anxious investors uncomfortable about financing a creative climate tech startup. Climate complacency can also have a more sinister explanation—the political influence still enjoyed by the old fossil fuel economy.

Ultimately, I believe that the big picture comes down to this: climate policy first and foremost must promote mitigation and adaptation. But catalyzing innovation, to ensure that decarbonization technologies will be waiting for us when we reach the next stage on this "net zero"—or, ultimately, "net negative"—journey, must also become a paramount objective. Notwithstanding an abundance of lip service about emission commitments, carbon neutrality by 2050, and intergenerational justice, public policies far too often are simply obtuse, oblivious to the full technological transition we need to restore the planet's climatic stability.

Yale University law professors Zachary Liscow and Quentin Karpilow have proposed that the UN Climate Change convention be amended so that countries are required to submit two separate action plans (that is, nationally determined contributions, or NDCs): the first, like present procedures, would describe intended national efforts to reduce greenhouse-gas emissions; the second would detail strategies to accelerate "cleantech innovation," including the anticipated initiatives of the local private sector. This is not a bad idea. Presumably, it would force governments to pay closer attention to the challenges faced by their respective climate tech industries and how governments might help through decarbonization interventions. At the very least, it would serve to promote a "clearer understanding of countries' contributions to the global cleantech knowledge stock."[1]

Initially, the imperative of promoting policies that remove obstacles and speed up climate tech development was not easy for me to internalize. As a lifelong environmental advocate, academic, and politician, I had always been quite content as a pro-regulation zealot: if *tough* rules designed to control an environmental challenge are not solving the problem, then it probably just requires *tougher* regulations. The climate crisis suggests that this approach does not always make sense. Instead, I was surprised to find myself berating government officials for being too stringent and inflexible in permitting or approving agrovoltaic project applications. If we are to dramatically reduce greenhouse emissions, we need to enact policies that simultaneously reduce our carbon footprint *and* nourish technological innovation.

Global warming constitutes a unique environmental challenge, a challenge unlike any other humanity has ever encountered. There are many reasons why it can be classified as a "wicked policy problem." The underlying social and scientific complexity means that there is no single, certain solution. Automatic support for draconian environmental regulations that delay climate tech deployment needs to be reconsidered because "the climate conundrum" really is different. The first thing that makes it different is that it simply won't wait. Dangerous thresholds will get crossed. As we *push* policies to rapidly reduce emissions, we must make sure not to push *away* the very innovations that might enable us to move toward carbon neutrality and climate stability.

Given limited resources, how should the government help? When asked, most climate tech entrepreneurs shrug their shoulders and say, "We'd be thrilled if the government would just get out of our way." But I believe that we, as a society, need to do more than simply stand aside and embrace some *laissez faire*, libertarian fantasy. History is replete with technological transformations and associated achievements that were only made possible by government intervention. Ambitious technology policies and proactive government engagement put a man on the moon and the Internet in our homes. We will not get to net zero without them.

The truth is that countries have been adopting policies that support decarbonization technologies for decades, and climate tech is responding. But the greatest challenges remain ahead. The International Energy Agency estimates that some 50 percent of future carbon dioxide (CO_2) emissions will need to come from technologies that are in the demonstration or prototype stages today. What can be done to shepherd them forward?

After considerable searching, I was surprised to find no comprehensive, accessible, and systematic review of policies that seek to mitigate climate change and to help climate tech entrepreneurs cross the proverbial "valley of death" through which they must pass to reach a marketable product. Nor is there a book that evaluates how disparate public policies can empower corporations to race ahead to develop and scale sustainable, low-carbon products. There is, of course, a surfeit of excellent publications about climate change. But no satisfying playbook exists describing the kinds of government programs that have begun to effectively cut emissions while bolstering climate tech ecosystems.

Such an evaluation requires a very different way of thinking about the role of government than the usual environmental policy manifesto. It seemed like a good idea to try a different path—to consider new approaches and hear new perspectives. Stanford's Graduate School of Business turned out to be an ideal place to undertake such a journey. I had written two previous books while on sabbaticals at this magnificent university. The chapters flowed quite naturally in the familiar embrace of the university's iconic conservation biology group. I was very much within my comfort zone as an environmental activist. But the climate crisis forces us all to change. Eventually, it was time to move on. Given the influential role that the business school has assumed at Stanford's new Doerr School of Sustainability, it made sense to join their team. And thus a new course in public policy and climate innovation was offered—and this book was born.

The book follows the progression of my Stanford course's classes and content fairly closely; it begins in chapter 1 with a cursory review of technology policy in general and the implications for the climate crisis in particular. Do we really want government intervention in the rough-and-tumble of climate tech start-ups? What do theory and empirical research teach us about nurturing innovation? In what ways is climate innovation different from innovation in other sectors, where governments seek to expedite technological breakthroughs? In chapter 2, we consider what sort of obligations countries take on pursuant to the United Nations Framework Convention on Climate Change and how international agreements influence domestic innovation policies.

As one looks around the world, it is possible to identify a rich suite of already-implemented policy alternatives that are supposed to support innovation and scale new climate technologies. Among the more common interventions are funding for research and development; carbon taxes and emissions trading (along with other economic subsidies); renewable portfolio standards, tax credits, feed-in tariffs, and carbon contracts for difference; technology-forcing prescriptions; government procurement programs and more gentle government nudges. They need to be compared. Which ones seem to work? What problems have they encountered in the real world? How can failed policies be fixed? And which alternatives make the most sense as we scramble to reduce discharges of greenhouse gases and galvanize a transition to a new economy?

The bulk of this book, chapters 3–8, considers how existing climate policies have fared, not just in reducing emissions but also in catalyzing technological innovation. My conclusions are taken from the vast academic climate-policy literature published in recent years, along with dozens of interviews of practitioners and observers in countries around the world. Many important lessons emerge from truly diverse international experiences. Rather than just muddling forward, there are numerous possible paths and priorities that need to be considered. The stakes are too high for us not to pick the best ones.

The final two chapters address two issues that are sometimes overlooked in discussions about climate tech, each of which reflects the growing recognition that climate justice, both domestically and at the international level, needs to inform public policy. The first issue concerns *disruption*, the phenomenon of workers losing their livelihoods in a changing economy. Of course, history is filled with Cassandras who offer dire predictions about the downsides of technological transformation and its many innocent victims. When automobiles and direct telephone dialing were first introduced, disaster scenarios were predicted for stable hands and switchboard operators. Yes, many people found themselves out of a job. But many more ultimately found a new and more prosperous niche in an expanding economy.

Similarly, it does not take a prophet to predict that the transition to a low-carbon economy will disrupt many people's lives. Indeed, there are already clear "losers." And many more will be left behind. If humanity reaches its carbon aspirations, then in less than thirty years there will probably be very few dairy farmers, coal miners, taxi drivers, or auto mechanics left. Climate advocates are quick to highlight the new professions and opportunities that a clean-energy transition will create. But it is still important for a compassionate, responsible society to support displaced workers through alternative employment programs and other measures to ensure a "just transition." Policy can help provide solutions for the many people who, through no fault of their own, are left without a livelihood.

The second chapter concerning climate justice tackles the "inconvenient truth" that the poorest places on earth, whose inhabitants have contributed the least greenhouse gasses, suffer disproportionately from the adverse impacts of climate change. This needs to be addressed in policies supporting adaptation projects. At the same time, it is absolutely essential to consider how policies can scale renewable technologies in developing economies. The United Nations projects that by the end of the century there will be four billion people living in Africa. That's a twenty-fold increase since 1950, when there were 220 million Africans. To stabilize greenhouse gasses in the earth's atmosphere, developing countries must be part of the solution. As countries begin imposing border taxes on goods with large carbon footprints, the Global South will need to be part of the transition if it is to fully participate in international markets. International leaders pursuing net-zero strategies have to do a better job of helping to deploy renewables and other climate technologies in the developing nations.

Unfortunately, environmental entrepreneurs tend to ignore developing economies because of their low market potential. That is one of the reasons that the air and water there are oftentimes so polluted. Climate tech must be different. Like most folks, I believe that it is high time we take the next giant stride forward in eliminating extreme poverty. But even those who do not believe in international development aid can understand that our *atmospheric commons* require that no country be left behind.

This book was written in an academic setting. But it is intended to address real-world problems—perhaps the most pressing global problem of them all. The different chapters can serve as a guide of sorts to scores of elected officials, civil servants, academics, business leaders, engineers, and activists in the trenches who are fully engaged in promoting policies to address the climate crisis. To support the many stories told, statistics shown, and arguments made, the book compensates for brevity with intensive documentation and footnotes. It reflects much of what I have learned from the most recent round of climate policies. What works? And what does not?

We are indeed living in perilous times, times that cry out for action. One third of Pakistan was underwater in 2022, leaving millions of homes fully submerged and tens of millions of lives ruined after unprecedented flooding. In 2023, thousands died in another apocalyptic, climate-triggered deluge, this time in Libya. Lytton, British Columbia, witnessed an unimaginable 49.6°C temperature in 2021 before burning to the ground. A single blaze, the Dixie Fire, just one of the 7,000 plus annual conflagrations that are now the new normal in California, decimated one million acres of forest. But that's only a fourteenth the destruction that occurred in Australia when massive bush fires raged out of control for months. As I write this, devastating fires raging across Canada and Greece have displaced thousands and sent suffocating smoke and particulates to faraway lands, closing airports and creating health hazards.

Siberia is sizzling again this summer, with very disturbing implications for the permafrost. UN experts estimate that the majority of the 60 million people displaced internally this year became refugees due to climate change. The oceans steadily rise, threatening the very existence of island nations like the Maldives or large swaths of low-lying countries like Bangladesh. The Micronesian island of Kiribati is only the first country whose citizens will have to move due to the rising sea.

The present carnage and chaos are but a trailer for the nightmare we can expect if the world fails to act boldly. It is truly difficult to believe that such a turbulent reality is real. As author Jonathan Safran Foer writes, we know the mounting climate disasters to be *true*. But at the same time, we are unable to *believe* them. It is simply too hard to wrap our minds around the implications of such a climatically violent future and what we must do to stop it. Nevertheless, the time has long been up for confining our efforts to education and explanations. Climate-driven catastrophes are affecting lives in virtually every corner of the world already. They share one disconcerting common denominator: the crisis is just getting started. If we fail to act, things will get much worse. The climate crisis demands far more effective public policies to reduce our emissions and stimulate climate innovation.

The good news is that if we do act, things can and will get better. If we muster the political will, there are already proven policies that can expedite many of the changes so urgently needed. If we design these policies to incentivize and harness human ingenuity, scientists and engineers can reduce—and ultimately eliminate—our collective carbon footprint. In order to transform carbon-neutrality objectives from wistful lip service to a new way of living on the planet, humanity must live by a new set of rules, and we must innovate. This won't happen by itself. But if governments intervene effectively, it can be done. This book describes how.

CHAPTER I

Tech Policy and the Climate Crisis

It was another oppressively sultry August day in Washington, DC, but the atmosphere in the White House was nothing short of jubilant. A last-minute compromise between the renegade Democratic West Virginia senator, Joe Manchin, and Senate Majority Leader Chuck Schumer delivered the deciding hand in the 51–50 vote in favor of what President Joe Biden called "the most significant US climate legislation ever."[1] Models suggested that the 2022 Inflation Reduction Act (IRA) would reduce the country's 2030 carbon footprint by 40 percent relative to 2005: a quantum leap in the direction of a "net-zero" emissions future. No matter that Manchin had insisted on framing the new statute as an anti-inflation measure and not one targeting the scourge of climate change. Everyone understood that the heart of the bill was an unprecedented $369-billion allocation for deploying climate tech and developing low-carbon technologies. The law's disparate tax credits for incentivizing decarbonization are likely to reach far higher sums.

America's return to climate leadership was a great relief for the growing number of people around the world alarmed that global warming was irreversibly damaging the Earth and threatening the future of civilization. The IRA offered validation for those who had long held the belief that the climate crisis could only be solved by rapidly decarbonizing the planet's economy. Such a dramatic transition would not happen without a massive investment in climate tech. The former vice president, Nobel peace laureate, and climate prophet-in-chief, Al Gore, tweeted about "a historic turning point": "It represents the single largest investment in climate solutions and climate justice in US history. Decades of tireless work by climate advocates across the country led to this moment."[2] Biden's former boss, Barack Obama, waxed optimistic: "Progress doesn't happen all at once, but it does happen—and this is what it looks like."[3]

At the same time, there was no shortage of critics who did not buy the underlying assumption of President Biden's visionary climate strategy. The Heritage Foundation, a leading conservative think tank, expressed the umbrage of many opponents:

> These subsidies could shift trillions of dollars of investment away from conventional energy sources and into green energy pipe dreams. This shift would leave our economy smaller, less dynamic, and less innovative, and will trap millions in poverty. . . . Do you like your natural gas stove or fireplace? Well, this bill is part of a broader effort to make these appliances relics of the past. It creates tax credits to have homes run on "clean energy" and for the purchase of "clean vehicles." If American consumers demand those types of products and features, that's one thing. The creation of this tax credit is a recognition that Americans don't desire the products and, therefore, Washington politicians must induce Americans to "do the right thing."[4]

A year later the debate still raged. Republican lawmakers proposed legislation to cancel the IRA before it had a chance to become fully operational.[5] Some politicians still assailed its policies as disgraceful handouts that subsidize elitist technologies for the wealthy.[6] Many American allies, from South Korea to Europe, were furious with the crass protectionism and preferential treatment that the new American climate strategy afforded the local climate tech industry,[7] claiming the policies threatened to set off a clean-energy "arms race."[8] Climate activists fretted that the IRA's support for decarbonization was "too little, too late,"[9] or simply wondered whether any government was capable of accelerating climate innovation fast enough.

The controversies beg two fundamental questions: First, should governments be in the "technology policy" business in the first place? And second, is technology policy an effective way for governments to reduce carbon emissions and confront the climate crisis? If the answers to these questions were not unequivocally affirmative, this book would be very short. Nonetheless, before diving into the clear imperative for government support for technological advancement in general, and decarbonization technologies in particular, it is worthwhile to consider some of the critics' common arguments.

There is no shortage of naysayers who attack the very notion of a systematic, vigorous technology policy, choreographed by a central government as ill-advised and wasteful. They have their reasons: for starters, they challenge the blind assumption that societal investment via government funded research in the private sector will ultimately pay for itself. Governments do not face the pressures to quickly produce favorable economic results that venture capitalists and banks do. This means that public funds may well be invested in ventures that will be less efficient and less profitable than their competitors that can already show lucrative "bottom lines" sufficiently appealing to attract investors. According to this view, government involvement in promoting technologies undermines the rough-and-tumble of free-market competition and assumes that taxpayers' money is less valuable than market-rate capital.

Ideologically, technology policy is anathema for capitalism's true believers. To them, flouting the natural "fair fight" of the free-market undermines the cherished survival-of-the-fittest dynamic on which a healthy economy is based. As they see it, unfettered competition has always been the best guarantor of optimal resource allocation, sound economic decisions, and technological progress. According to this view, the only legitimate role that governments should assume in supporting innovation is ensuring that markets remain open, transparent, and competitive in order to allow "the best technology to win." In the course of human events, market distortions may occur for any number of reasons. But governments should try to minimize them, not perpetuate or exacerbate them through policy that promotes certain technologies. Making the advancement of innovation an integral component of public policy leads to swollen bureaucracies, with all the institutional ambition and hubris that inevitably spawn partiality and wasteful economic decisions.

Another argument against technology policy goes like this: In today's global economy, progress depends on the removal of barriers and the free exchange of ideas. In a post-Covid world where an increasingly high percentage of researchers and developers work remotely, the notion of a country devoting resources to gain a national economic advantage through a particular technological advancement or domination of a manufacturing sector is simply no longer appropriate. A truly *global* economy requires an open, creative, transboundary, and ultimately unobstructed competitive process.

Competence is another source of objections: critics argue that governments are not designed—and indeed lack the capacity—to pick technological "winners and losers." While many talented officials may harbor delusions that they can shape the outcomes of technology development, or that strategic investments in good ideas invariably lead to prosperity, the record suggests otherwise. The technological stagnation that characterized the centralized Communist economies of the past are a reminder of the perils of bureaucratic overreach. Especially in free-market economies, a ponderous government hand at the steering wheel, guiding technological strategy, is unlikely to lead to favorable outcomes.

In the context of climate tech, the case of Solyndra offers a case in point: this Fremont, California–based company offered tubular solar panels that were supposed to be uniquely appropriate for large, low-sloping roofs. Their cylindrical systems were made from copper indium gallium selenide, which the company argued would be more efficient than conventional crystalline silicon. In 2009, as part of the renewable energy boost at the heart of the Obama Administration's economic stimulus plan, the government cosigned $535 million in loans for the company. There was no shortage of warning signs: the founding CEO had left the company; market analysts were wary; Dun & Bradstreet's credit appraisal of the company was a very diffident "Fair."[10] And of course, China was subsidizing its solar manufacturing so much that competing was practically impossible.[11] But the government was on a mission, intent on making a significant financial commitment to renewable energy. Disaster did not take long. Two years later, Solyndra filed for bankruptcy.[12]

The Solyndra debacle raises the political problem of "sunk costs" and bureaucratic momentum. Civil service culture and institutional exigencies don't allow officials to admit when they make mistakes in their use of taxpayers' money. Time after time, governments throw "good money after bad" at projects: from dams and supersonic jets to nuclear plants and synthetic fuels. Conversely, long investment horizons don't jibe well with election cycles: shifting political winds in democratic societies inevitably truncate funding and lead to the collapse of long-term projects.

There are also legitimate concerns about the unintended consequence of crowding out private financial markets. Taxpayer dollars are precious and public coffers limited, making government budgeting a zero-sum game. Tax-funded programs to advance new technologies (when the same innovation might have enjoyed support from the private sector) may come at the expense of underwriting other important civic or social initiatives. Venture capital (VC) will never be able to compete with the government in assuming risks of unproven technologies, taking on long-term investments, or absorbing the losses when an R&D venture goes south. Investors may play it safe and simply opt out, whereas in the absence of government investment, some might have been very willing to step up to the plate and finance a risky new technology.

Technology policy is also disparaged by critics for spawning corruption and favoritism. No bureaucrat is born fully dispassionate and unbiased. Very gifted, persuasive, and well-connected people are paid large sums of money to influence public spending decisions. Even if no money is passed under or over the table, far too often enticements, whether political or material, inform government decisions—generally, not for the better. Even if one holds a less cynical view about the way lobbyists operate, access to decision-makers will never be created equal.

Opponents of government involvement in directing technological development invariably offer "empirical proof" to support their position. American conservative think tanks are particularly good at finding examples of confident bureaucrats wasting unseemly sums of public money to promote technologies that, in retrospect, were always duds. Take the time the city of San Francisco spent over $20,000 on a trash can.[13] And there is the $400,000 self-cleaning toilet for the Washington Metro that never really worked.[14] NASA's exotic inventions for life in space are living proof that the government is not always a paragon of savvy technology development. Should it really cost $125,000 to develop a 3D "pizza printer" that might be used on Mars fifteen years in the future?[15] And of course, the veil of secrecy behind military technologies hides the most grotesque waste of public resources for new weapons system. Consider the $1.7 *trillion* dollars spent since 1995 on the F-35 Joint Strike Fighter jet, which is still not ready to fly.[16]

So go the main objections of technology policy detractors. They may make for good rhetoric. But this jaundiced perspective about government involvement in technological innovation has never had much traction historically. In recent years it has enjoyed less and less influence and is largely dismissed as demagoguery. There are good reasons for this.

The Innovation Imperative

Technology development has been an important part of government activity around the world from time immemorial. Indeed, public commitment to support promising technologies is neither radical nor new. It was America's good fortune that Alexander Hamilton emerged as the dominant voice in economic matters in the formative days of the United States. As early as 1791, Hamilton was outspoken about the need for the government to cast aside *laissez faire* orthodoxy for a range of policies—from tariffs to procurement contracts.[17] In Hamilton's hyperactive mind, government intervention was essential to foster industry in the nascent national economy and make the United States competitive with Europe.

The American government began to allocate funds for technological advancements in weaponry almost from the get-go, starting with the establishment of its Army Corp of Engineers in 1802.[18] Years later, Congress established the Office of Science and Technology Policy in its Department of Industry and Commerce to make sure that the president has the best possible guidance on advances in science and technology in *civilian* matters. This office continues in its mission to "advance health, prosperity, security, environmental quality, and justice for all Americans."[19] On average, the US government spends more than $70 billion a year to fund scientific research and nurture innovative technology. Of course, this is hardly an American phenomenon. A sustained institutional commitment to supporting technological development has been part of national economic strategies in Japan, Israel, and South Korea, as well as across the European Union from Germany and Ireland to Denmark and Sweden—to name just a few. Apparently, these programs enjoy sufficient demonstratable successes to prevail over libertarian skeptics.[20]

Investment of public funds in technological advancement is sound public policy. There are three central justifications for sustained government commitment to technological innovation:

- the critical role of governments in ensuring economic prosperity;
- the responsibility of government to provide public goods; and
- market failures and the need for government intervention to overcome "externalities"—the negative side effects of commercial activity that are not reflected in the cost of products and services.

Economic Prosperity

Economies are driven by four factors that affect the production of goods: natural resources, physical capital, human capital, and technological innovation.[21] Improving these "inputs" leads to higher GDP and greater total prosperity. It should therefore not be surprising that governments take an active role in nurturing *all four* factors. There is a broad societal consensus that allocating tax dollars toward the sustainable development of land, water, and minerals (natural resources), infrastructure (physical capital), and education (human capital) is an appropriate use of public funds. Innovation is no different. Insofar as technological innovation generates economic activity, and economic activity generates social prosperity, investments in expanding a society's technological tool kit is a sound, if not obligatory, use of public funds.

National agencies and programs with a mandate for proactive technology patronage make sense as the relentless velocity of change grows ever swifter, making it imperative for businesses to run ever faster in order to stay in the global game. Even large, seemingly well-managed corporations are not always sufficiently nimble to remain competitive. The decline of Nokia, once the world's preeminent cell phone company, is instructive. By 2010, the Finnish company's market share had begun to hemorrhage as Apple and Samsung came to dominate the smartphone market.[22] There are several reasons for its economic implosion, but chief among them was Nokia's inability to stay apace of the rapidly evolving technologies in the cell phone industry.[23]

While much is made of the global nature of today's economy, there are other forces at play. In a post-Covid world, many countries are wary of becoming too dependent on international markets and are beginning to hedge their bets with homegrown alternatives. Governments are looking to repatriate supply chains and localize innovation. In most economies, this will not happen without policy interventions.

Similarly, if a national government is expected to help sustain its country's competitive advantage, it must ensure that its industries constantly innovate.[24] Wise policy, for instance, played a key but often underappreciated role in supporting the technology advances that led to the rise of the "Asian Tigers"—the booming economies of South Korea, Singapore, Taiwan, and Hong Kong—in the second half of the twentieth century.[25]

Regardless of a nation's particular circumstances, innovation remains one of the most reliable indicators of economic and societal health. Indeed, there are few universal truths as salient as the need to evolve and adapt. Evolution theory was predicated on the axiom that it is "not the strongest of the species that survives, nor the most intelligent that survives. It is the one that is most adaptable to change." This is as true for economies as it is for biodiversity. From Charles Darwin to Jared Diamond, the lesson is plain: Adapt or die.

Providing Public Goods and National Security

National defense is a classic example of a public good because all citizens benefit from it: there really is no way to exclude "non-payers" and prevent "free riders" from enjoying military protection. The same holds true for law enforcement, highways, education, or clean air. Sadly, in a world that favors "the strong," a country's survival is often more a function of military might than economic agility. That's why, historically, the lion's share of government spending on technological innovation goes to ensuring national security. Indeed, the United States spends some $773 billion dollars annually on its defense budget,[26] with a large portion directed to applied scientific research and innovative military technologies.

Aside from providing security, military investments can also yield extraordinary technological advances with applications to civilian life. Afterall, the Internet, nuclear energy, speech recognition software, autonomous vehicles, LED lighting, magnetic resonance imaging (MRIs), and even freeze-dried foods all started as military projects. But even without side benefits to society, the imperative of keeping citizens safe will always drive government investment in R&D.

Addressing Market Failures

The invisible hand of the market does many things well, but it cannot guarantee that the economy will benefit society as whole. For example, knowledge is essentially a public good. Typically, for an inventor to be remunerated for her work, her breakthrough cannot be kept a secret. For most scientists to succeed, their research findings also must be made public. This in turn can create a virtuous cycle or spillover effect, in which other researchers build on a new discovery to create additional innovations. The public surely benefits from this dynamic, but the inventor, or the company that invested in the research, may not. As a result, research to develop a critical technology may be underfunded, producing a less than optimal social outcome.[27]

There are many instances when it becomes the government's responsibility to intervene in order to correct market failures. "Orphan drugs" constitute a classic example. In many cases, narrow market analysis leads pharmaceutical companies to abandon development of a drug that could cure serious but rare diseases. Typically, this happens when fewer than 200,000 people are affected. That's why the US Congress passed the 1983 Orphan Drug Act, guaranteeing the pharmaceutical industry sufficient financial benefits to develop innovative medical solutions in such cases.[28] In retrospect, due to this compassionate and rational policy, some 600 orphan medicines for rare diseases have come to market during the past forty years.[29] Because the government stepped in when the market wasn't working, millions of people around the world were given the gift of life.

Much of environmental protection policy is predicated on the notion that economists have a hard time ascertaining the full social cost of polluting activities. The truth is that environmental assets in most transactions are undervalued. This is the case for clean air, healthy oceans, safe drinking water, quiet, uncontaminated land, biodegradable products, and endangered species. As a result, no one is willing to pay for the technologies that should be developed to prevent adverse environmental outcomes—even though the collective societal benefits from doing so greatly outweigh any associated expense.

In cases of market failure, there is a compelling public interest in government involvement to promote a technological solution. And so it was that myriad unaddressed environmental problems gave rise to interventions that spawned key technologies, including biowaste treatment developed by the UK government under its Landfill Directive;[30] new German biodiesel production;[31] wastewater treatment technologies in France;[32] and advanced oxidation processes for treating contaminated ground and surface water, developed by the US federal government in its hazardous waste remediation Superfund program.[33] The damages associated with the climate crisis are orders of magnitude higher. In 2006, London School of Economics professor Nicholas Stern estimated that the total costs of climate change would soon reach 5 percent of annual global gross domestic product; as conditions worsen, the global costs might reach an unimaginable 20 percent of GDP.[34] No subsequent report has changed the persuasive economic rationale for investing in climate tech.

The public has come to rely on the government to ensure that society develops the technology it needs to remain competitive economically and to defend itself. It also expects it to address the many market failures that lead to abandonment of essential technologies. Support for scientific and technological advancement constitutes a critical government function and should be fully reflected in the global strategy to address the climate crisis.

Why Climate Tech Policy Is Different

Promotion of climate tech has many characteristics in common with other efforts to advance technology, but it also has inherently unique dynamics. What then makes climate technology different from other areas of economic development? Climate tech has four unique traits, any one of which justifies particularly creative and aggressive policies for stimulating innovation: externalities, extended development horizons, high costs of early experimentation, and urgency.

Externalities

Climate tech's sluggish initial performance in the late twentieth century constitutes a classic case of "market failure"—when the value of a business venture to society is far higher than its current commercial value, and governments do not intervene. This is a function of externalities, or the "costs and benefits that are not captured in market prices."[35] This divergence is particularly pronounced when it comes to climate tech, which means that sustainable technologies may not be funded by private capital, even when there are powerful societal benefits that justify investment. For example, for years developing photovoltaics and storage made sense ecologically—and economically—in countries importing fossil fuels, but it never happened. Like so many other climate-mitigation challenges, without a price on the greenhouse gases released to the atmosphere, climate tech development will never be sufficiently monetized and will always be undervalued.

It is worth remembering that society has no problem penalizing firms that produce *negative* externalities. The "polluter pays" principal is axiomatic across the world as the basis for civil and criminal penalties in environmental legislation.[36] By the same token, subsidies are appropriate when technologies create *positive* externalities—like the collective benefit gleaned from transitioning to renewable electricity, energy storage, cultured protein, carbon removal, or low-carbon cement.

There are economic formalists who a priori oppose any imposed "internalization of externalities" because they see it as a violation of free-market capitalism. But almost without exception, modern governments accept the notion that the "invisible hand of the market" frequently does not produce optimal results. It is well to quote the late Herman Daily, the World Bank economist turned University of Maryland professor, who quipped, "If the survival of humanity is external to your model, you probably need a new model." Policies that prioritize support for climate tech are also an antidote to the many years of subsidies lavished on the fossil fuel industry with its negative externalities[38]—and an expression of the positive externalities associated with technologies that mitigate climate change.

Extended Development Horizons

Unlike conventional high tech, where developing an app or cyber security software can reach the market in a matter of a year, climate technology takes time. Climate tech is often tightly linked to infrastructure—and infrastructure planning and construction are protracted processes. The problem is that venture capital funds generally are set to run for no more than ten years—and usually far less.[39] At that point, any remaining funds are liquidated, and financiers remunerated. Climate tech entrepreneurs require more "patient investors."[40] While politicians operate according to short time horizons, government bureaucracies staffed by career civil servants have the ability to take a long-term view.

"Green hydrogen"—that is, hydrogen produced by the electrolysis of water using renewable electricity—offers a good example of just how long this proverbial "valley of death" can be for firms that seek to develop meaningful climate tech innovation. *H2Pro* is a promising Israeli green hydrogen firm that opened its doors in 2019.[41] But the company's technology is based on five previous years of electrochemistry research by Techion research team Avner Rothschild, Gideon Grader, and Hen Dotan. The team developed an electrochemical thermally activated chemical (ETAC) method for splitting water, a breakthrough that was important enough to be published in *Nature* in 2019. The company's innovative design is promising indeed: a typical hydrogen electrolyzer is only about 70 percent efficient; 30 percent of the energy used in production is lost as heat. Unlike conventional electrolysis, H2Pro's process generates hydrogen and oxygen separately in different steps—an electrochemical step and a thermally activated chemical step. This allows for high-pressure production of green hydrogen in a way that retains exceptional energy efficiency (98.7 percent HHV) inside the reactors.[42]

Today, the team has grown from Rothschild's original four laboratory staffers to 100 workers, and the company is planning to build a factory. All this requires a significant budget, even though H2Pro has yet to produce its first demonstration unit and cannot know for certain whether its unorthodox process will work on a commercial scale. And even if it succeeds in cost-effectively producing green hydrogen at scale, profits will take at least a decade from initial investment to materialize.

Tech investors like to look closely at a product's "technology readiness level." The TRL system involves a nine-step progression, developed by NASA during the 1970s, that is useful for assessing the maturity of technologies for acquisition. It starts with basic science and feasibility research (levels 1 and 2) until ultimately reaching system development and system testing (levels 8 and 9). Professor Rothschild characterizes the TRL sequence economically as "essentially a logarithmic scale."[43] In other words, as a company progresses, moving up through the different technology readiness stages, the costs rise.

The other factor that makes development horizons so long is the need to scale technologies so that prices can drop to economically feasible levels. For a climate tech innovation to be successfully diffused, a lower-cost alternative typically must be available to replace an existing technology. H2Pro is the exception that proves the rule. It succeeded in raising over $100 million dollars from investors including Bill Gates' Breakthrough Energy Ventures and Israeli tech entrepreneur extraordinaire, Talmon Marco. This gave the company the necessary capital to build a commercial-scale facility for producing hydrogen. Presumably, idealistic billionaires have the stamina and the monetary muscle to see this project through.[44] But it is unlikely that conventional VC investors can wait so long or are willing to assume the associated financial risk.

Driving the price of a disruptive technology down to a reasonable level does not happen overnight. But when it happens, it can happen fast. "Wright's Law" was developed to quantify the pace of this process. It was in 1936 that aeronautical engineer Theodore Wright observed a drop in the production costs of airplanes during the preceding decades. Each year, the airframes manufactured became less expensive. As he formulated a historic cost curve, a clear pattern emerged: for every doubling of production, Wright saw a steady 15 percent reduction in costs.[45]

The same trend holds for climate technology: "Swanson's Law" is the renewable energy equivalent, named for Richard Swanson, the Stanford University engineering professor and founder of SunPower, a solar photovoltaic cell manufacturer. The first solar cell was invented in the United States in 1954. But a year later, Japan made a prototype model of a solar cell and was quick to take the lead in developing the technology. By 1958, Japan had installed its first photovoltaic system, with a generating capacity of 70 watts, at a radio relay station on top of Mount Shinobu. It took another twenty years before photovoltaic electricity systems were connected to the country's power grid: in 1978, Sanyo Electric began installing PV generation systems with the capacity to return the surplus electricity from individual houses to the power supply system. This breakthrough was soon reflected in expanded production. Meanwhile, PV cells were being used by Japanese manufacturers to power calculators. By 1999, Japan was the world leader in the total manufacturing of photovoltaic solar cells and, for a brief period, in PV electricity generation.[46]

Professor Swanson was among the first to demonstrate that solar electricity prices dropped by 20 percent for every doubling in solar panel capacity.[47] Tracking the more recent lowering of prices of solar and wind energy suggests that the price decline has been even more precipitous, reaching as much as 30 percent with every 100 percent increase of installed base.[48] As a result, by 2022 solar energy was 30 percent less expensive than electricity produced by natural gas.[49]

This relative advantage was the result of some forty years of constant multiplying in production.[50] Expanded production and massive deployment of decarbonizing technologies also yields innovation, which is soon manifested in lower costs. This is why it is impossible to separate general climate-mitigation policies from climate innovation. Programs to reduce greenhouse-gas emissions through deployment of clean technologies arguably constitute the single most powerful driver of climate tech R&D.

Moreover, for private capital to gain a reasonable return on investment, innovation has to have a large enough technological "moat" to ensure exclusivity. Many solutions, even if viable and marketable, are simple and easy to emulate. This makes them poor candidates for venture capital. Take, for instance, meat alternatives, which are much less carbon-intense than animal protein. One of the reasons why many analysts believe that Impossible Burger will ultimately beat Beyond Meat in the battle for preeminence in the plant-based protein niche is that the latter is based on technology which is relatively easy to copy.[51]

All this means is that climate tech promotion is simply a great place for public policy to shine. Government involvement is critical for providing the oxygen for completing the many long-distance runs that will be needed to reach the finish line in humanity's race to a sustainable economy. If designed correctly, policies can provide numerous ways for making VC investment more compelling.

The High Cost of Early Experimentation

It is estimated that only about 1,000 of the 500,000 new startups each year in the United States succeed in receiving first-round venture capital funding.[52] That's because startups in general are a risky business; with its high incubation requirements, climate tech is even more so. As part of a *derisking* strategy, many venture capitalists like to frame technology startups as early-stage experiments. Presumably, the aspiring company constitutes a hypothesis. Like any theory, it can be tested through an iterative experimentation process, with benchmarks of success or failure along the way.[53] For instance, whether a protein substitute will appeal to consumers can be tested for a lot less than the cost of building a factory. As the product proves its marketability, financing becomes easier. Still, meeting meaningful benchmarks can be costly.

The case of Quidnet Energy is instructive. The California-based startup has developed a unique geomechanically pumped energy-storage system. Given the cost and the solid-waste challenge posed by lithium and other battery technologies, the Quidnet system offers a promising response to the intermittency of renewables. The idea, at least in theory, is simple: when solar or wind systems produce a surfeit of electricity, the system pumps water from a pond, down a well, into a body of rock. As soon as electricity is required, the well is then opened so that the pressurized water can pass through a turbine to generate electricity. The water then returns to the pond for the next cycle.[54]

Quidnet was established in 2012, but like so many climate tech startups, it had difficulty attracting venture capital. By 2016, it was stuck. For the company to move forward to the next stage in its development, a field trial was required for proof of concept. There was no guarantee that the test would succeed. Company engineers saw it as a binary, up-front risk: immediately upon its completion, they would know whether their system worked . . . or not. Unfortunately, the cost of the decisive test came to roughly a million dollars. Not surprisingly, no one was volunteering to pay for it. Ultimately, civil society became engaged. Prime, a newly formed NGO created precisely for this purpose, raised money from two foundations to conduct the test.[55] Using retired Texas oil wells, the trial was conducted and proved successful, allowing the company to proceed with subsequent seed rounds.

Private philanthropy can certainly play a role in combating climate change. Yet early-stage experiments designed to reduce uncertainty and allow startups to move to the next stage in their development are costly, demanding resources far beyond most foundations' capacity. A Harvard University Business School case study about entrepreneurial experimentation concluded, "The very long time-frames and costs associated with learning about a new way to produce clean energy, or different approaches to curing cancer, create a dearth of experimentation, despite intense societal interest."[56] Government involvement to help support and validate the testing is essential for reducing uncertainty and catalyzing investment from the naturally cautious private sector.

Urgency

As society confronts the climate crisis, momentum matters. Yale law professors Zachary Liscow and Quentin Karpilow describe its importance in terms of a process they call "innovation snowballing." This means that innovation builds on itself over time; present innovations will make future ones even more effective and more valuable.[57]

If smart decisions in climate policy are made today, presumably the next generation will be in a better place to realize the next technological phase when it reaches the gun lap in humanity's race to net zero—when a balance is reached between the greenhouse-gas emissions released into the atmosphere and those taken out. If, however, today's generation fails to act, the potential to develop the technologies necessary for a sustainable planet in the future will be diminished.

Unfortunately, talking about the future has proven to be lousy strategy for galvanizing a bold public response to the climate crisis. At the individual level, people see their children and intuitively understand who will benefit from their efforts to decarbonize. But at an amorphous, societal level, "intergenerational justice" is a much harder sell. To inspire greater commitment to combat climate change, advocates often employ military metaphors.[58] Societies at war presumably offer proof positive that if people would only understand what's at stake, they would make the necessary individual sacrifices (and governments would make the investment) required to arrive at an economy with net-zero carbon emissions— even far earlier than the 2050 target. If Americans were willing to plant Victory Gardens during World War II, surely today they should be willing to install double-glazed windows in their homes and photovoltaic systems on their roofs.

This rhetoric resonates nicely but has not proven to be persuasive. There are many reasons why. A decade ago, Canadian professor Robert Gifford identified twenty-seven psychological barriers that explain why people in general—and presumably decision-makers as well—do *not* feel compelled to act to decarbon- ize, even in the face of mounting evidence that immediate and radical action is necessary: mistrust, behavioral momentum, limited cognition, social compari- sons—the list of explanations goes on and on.[59] Each barrier constitutes another reason why governments are going to have to step up and accelerate the speed of climate tech development.

If the public and its representatives would only listen to scientists, presum- ably commitment to development of climate technology would increase dramati- cally. This may be starting to happen, but it certainly is not happening fast enough. In 2018, the United Nations Intergovernmental Panel on Climate Change (IPCC), a generally cautious consortium of independent climate experts, released a special report in which its usual vagueness was replaced with refreshing, if alarm- ing, specificity: in order to avoid a temperature increase exceeding 1.5°C, the world needed to reach an interim 45 percent reduction in global carbon dioxide (CO_2) emissions within twelve years.[60] Unfortunately, since that time, release of CO_2 has actually continued to rise, which makes achieving this goal increasingly unlikely. Many of the key technologies required to move the planet toward such targets still need to be developed, but they lack funding.

Whether because of complacency, myopia, or selfishness, humanity has not internalized a true sense of urgency about the crisis. Neither has the international community. Representatives of 200 countries convening at the 2023 UN climate conference in Dubai (COP28) resisted demands from leading scientists and cli- mate activists to begin a phase-out of oil, coal and natural gas, and merely calling for the "phasing out of inefficient fossil fuel *subsidies* that encourage wasteful consumption."[61] (Even so in contrast to the earlier gatherings, the Dubai decision recognized that global greenhouse gas emissions needed to peak "at the very latest" by 2025 if global warming was to be limited to 1.5 degree C.[62]) The ability of individual countries to influence the international agenda and United Nations Framework Convention on Climate Change (UNFCCC) targets appears to be limited, but there is no reason why those countries cannot support climate tech ventures within, or even beyond, their own borders.

Quite simply, unlike the other problems the world faces that require technological solutions, for the foreseeable future global warming appears to be irreversible. This makes the task of developing climate tech far more urgent. Innovation experts generally estimate that technologies discovered today are likely to have their biggest impact two to three decades later. Unfortunately, humanity is in a race against time. Experts warn that if global warming continues, before long the planet will cross key tipping points that will set off a cascade of catastrophic consequences.[63] To stop global warming, technologies that can scale far faster must be found.

Martin Luther King's call to arms from sixty years ago is entirely relevant to the present situation: "We are now faced with the fact that tomorrow is today. We are confronted with the fierce urgency of now. In this unfolding conundrum of life and history there is such a thing as being too late."[64]

Public Policies to Promote Climate Tech Innovation— What's on the Menu?

If government is justified in advancing climate tech, the question becomes: What innovations should it back and how should it support them? Just as the technologies required to move to net zero are hardly homogeneous, innovation itself can be broken into at least three different categories.

Incremental innovation improves existing technologies to make them more efficient without fundamentally changing underlying processes. Like a new model of car or cellular telephone, the new products are just a little better or a little less expensive than the previous version of the product or service. *Disruptive innovation* refers to more-profound changes in specific technological functions, without modifying the underlying system or technological regime.[65] The new product may create a completely new market. The transition from typewriters to word processors or incandescent to LED lighting are good examples. Finally, there is *radical* or *transformational innovation*. This phenomenon is relatively rare. Technologies central to an economy experience a fundamental shift such as the move from steam power to the internal combustion engine, or the Internet, which changed the entire nature of commerce and communications, or the transition that has already begun from internal combustion to electrically powered vehicles.[66]

It is not surprising, therefore, that different policy interventions are required to stimulate different kinds of climate technology transitions. Moreover, the needs of small companies and startups are fundamentally different from those of well-established industries. In the area of climate mitigation, there is no "one size fits all" policy alternative. Rather, a broad range of instruments needs to be applied to address the many pathologies that contribute to the climate conundrum.

One way to think about the various necessary approaches is to differentiate between policies designed to develop new technologies versus those designed to deploy existing technologies and make them affordable. It is critical that financial support for climate tech development be provided alongside incentives to meaningfully expand the *supply* of innovative low-carbon alternatives. An entirely different array of policies is required to increase *demand* for them.[67]

On the technology *supply* side are policies like the following:

- Direct government funding and subsidies for research and development;
- Tax credits and economic incentives to support the activities of climate tech companies in different stages of development;
- Patenting and protection of intellectual property for climate mitigation;
- Facilitating cooperative relationships between researchers and the private sector, including the establishment of technology incubators.

On the *demand* side are programs such as

- Carbon taxes and greenhouse-gas emissions trading systems;
- Tax exemptions and tariffs to incentivize decarbonizing technologies;
- Government procurement directives that prioritize purchase of climate tech products;
- Technology-forcing prescriptions that require adoption of a low-carbon technology;
- Carbon contracts for differences (described in chapter 5), in which governments change the economic calculus by covering the cost differentials associated with relatively expensive but promising new technologies; and
- Softer policy interventions, such as green labeling or disclosure requirements, that can gently nudge industries toward innovation and better climate performance.

The following chapters consider these alternatives and their design: what they are intended to do theoretically and what they actually do. While it is true that the "devil is in the details"—excessive detail and nuance can be devilishly numbing. So, in the following pages, the focus is on the big picture, basic principles, approaches, and outcomes from disparate programs. (Conscientious readers who wish to wonk out and take a deeper dive will have an easy time doing so by following the references.)

There are innumerable climate tech stories that demonstrate the soundness of different policy approaches. Because it constitutes the largest single source of greenhouse gases, decarbonizing electricity justifiably was the first climate challenge embraced by most countries. But it is only Act One in the larger drama that needs to unfold as humanity strives to save its planet. This book also considers how policies can support innovation in sectors that are proving hard to abate but are critical to achieving a net-zero economy.

Climate tech maven and podcast host Shayle Kann calls these sources "the eight horsemen of the Apocalypse": steel, aluminum, cement, chemicals, aviation, shipping, trucking, and chemicals. Together, they constitute roughly a third of total global emissions. The food system contributes another 25 percent; transportation probably 24 percent. Without effective government interventions, their relative contribution to the greenhouse-gas burden will surely grow in the coming years as emissions from electricity continue to drop.

But trend is not destiny. Sound policies can harness the world's abundant competence, creativity, and growing commitment to climate stability in order to hasten innovative solutions.

CHAPTER 2

A Global Framework for Mitigation

Few climate activists recognize the late British prime minister, Margaret Thatcher, as a champion of the cause, but in hindsight, it is fair to say that no world leader did more to bring early attention to global warming than the "Iron Lady." Thatcher was an unlikely figure to play a critical role in international climate politics. Throughout her tenure in office, she showed a visceral dislike for excessive regulation, opposed environmental initiatives of the European Community, and was an avid supporter of nuclear energy.[1] But as a young woman at Oxford, Thatcher had studied chemistry, and she retained an appreciation for science and scientists. When efforts to reign in chlorofluorocarbons (CFCs) and other ozone-depleting chemicals stalled, the prime minister used her formidable influence to upgrade international commitments and accelerate phaseout.[2]

This was about the same time that scientists were reaching the conclusion that the climate was changing. It is ironic that it took some time for the international community to get used to the notion that the world was getting warmer, not colder. During the 1970s, many serious scientists believed that pollution was leading to a new ice age.[3] But after British experts briefed Thatcher about the dangers of rising temperatures, she concluded that the issue could no longer wait.[4]

Every year, world leaders convene in New York to address the United Nations General Assembly. In 1989, after ten years at the helm of the United Kingdom,

Thatcher astonished the attendees at the UN plenary when, out of the blue, she devoted her entire address to a crisis that was, at the time, largely unknown. After presenting some of the evidence about "the greenhouse effect," she called for action:

> The most pressing task which faces us at the international level is to negotiate a framework convention on climate change—a sort of good conduct guide for all nations. Fortunately we have a model in the action already taken to protect the ozone layer . . . that aims to prevent rather than just cure a global environmental problem. . . . But a framework is not enough. It will need to be filled out with specific undertakings, or protocols in diplomatic language, on the different aspects of climate change.
>
> These protocols must be binding and there must be effective regimes to supervise and monitor their application. Otherwise, those nations which accept and abide by environmental agreements, thus adding to their industrial costs, will lose out competitively to those who do not. The negotiation of some of these protocols will undoubtedly be difficult. And no issue will be more contentious than the need to control emissions of carbon dioxide, the major contributor—apart from water vapour—to the greenhouse effect. We can't just do nothing.[5]

By the time the United Nations "Earth Summit" convened at Rio de Janeiro three years later, Thatcher had unceremoniously resigned, but her words proved prescient: a draft Framework Convention for Climate Change had already been negotiated and prepared for adoption. Delegates had been aware of disconcerting reports by the newly created Intergovernmental Panel on Climate Change (IPCC). This consortium of leading climate scientists had been formed in 1988 by the World Meteorological Organization, along with the United Nations Environment Programme, and was later endorsed by the United Nations General Assembly.[6] Its avowed purpose was to "provide policymakers with regular scientific assessment on the current state of knowledge about climate change."[7]

Treaty negotiations were acrimonious: developing countries were naturally disinclined to take action for a problem caused by the fossil fuel economies of wealthy Western countries. Eventually a compromise was reached, whereby the convention recognized countries' "common but differentiated responsibilities" in addressing the problem. This basically set the tone for the next three decades, during which the international community began to muddle forward toward decarbonization, while the global South was exempted from concrete actions to control emissions.

Thatcher and the original United Nations Framework Convention on Climate Change (UNFCCC) leadership could not help but be informed by the concurrent efforts of the international community to address stratospheric ozone depletion. A group of roughly a hundred chemicals was destroying the protective layer that shielded the Earth from the sun's ultraviolet radiation. Fortunately, relatively innocuous chemical substitutes for most of the pernicious CFCs were already being developed. It made a lot of sense for countries to simply adopt a "top-down" approach and force industry to use these substitutes to replace the halogenated compounds that were creating the "ozone hole." As far as international ecological challenges go, reversing stratospheric ozone depletion was looking like the ever-elusive environmental "quick win."

And so it was that the 1992 UNFCCC came to emulate the ozone-depletion agreement's top-down approach. All that was required was for the world's governments to prepare national programs to mitigate climate change by replacing sundry sources of anthropogenic greenhouse-gas emissions. The first generation of climate diplomats quickly learned, however, that climate change was thornier than the world's other ecological challenges. After all, there was no getting around the fact that the entire global economy was based on fossil fuels. Cutting emissions would require the development and widespread adoption of new technologies. Indeed, the agreement brokered at the 1992 summit, the United Nations Framework Convention on Climate Change, specifically calls for signatories to "Promote and cooperate in the development, application, and diffusion, including transfer of technologies, practices, and processes that control, reduce, or prevent anthropogenic emissions of greenhouse gases . . . in all relevant sectors, including the energy, transport, industry, agriculture, forestry, and waste-management sectors."[8]

In retrospect, the treaty's implicit assumption—that the global community could compel countries to both come up with climate-friendly technologies and then integrate them into every aspect of their economies—was naïve. Extremely naïve.

Given the difficulty of mandating this type of sweeping change, it is important to ask whether (and how) the international community—and in particular the United Nations climate accord—have actually influenced the development and adoption of technologies as well as policies to reduce emissions. To what extent does international coordination inform the scientific research agendas required for new advances? And what more should global cooperative initiatives do to accelerate climate innovation and ambitious climate policy?

A Global Crisis

One of the most significant barriers to the global adoption of climate tech is that, currently, it pays to pollute. Countries that are complacent about reducing greenhouse-gas emissions can produce goods more cheaply than their more conscientious counterparts, which means they have a leg up in international trade. It also means that over time, more and more products will be manufactured in countries with lax standards— reflecting a proverbial "race to the bottom."

In fact, as Oxford professor Dieter Helm points out in his book *Net Zero: How We Stop Causing Climate Change*, national governments' decarbonization efforts can end up being counterproductive if they fail to consider the global context. For instance, if a country with strict manufacturing regulations closes down a domestic factory to reduce its own emissions, it may well result in citizens buying more products from companies based in nations with more lenient regulations, leading to an end result of more total emissions. Decision-makers would do well to remember that *net-zero consumption* ultimately matters far more than *net-zero carbon production.*[9] This can only be achieved by policies that go beyond regulation of local sources of greenhouse gases.

Fortunately, countries are increasingly cognizant of these dynamics and are developing policies to eliminate the phenomenon, commonly known as "leakage." This refers to situations when businesses transfer production to countries with less stringent emission constraints. Border taxes (or as the European Union euphemistically calls them, "adjustments") are already on their way. These taxes extend the EU's domestic price on carbon emissions to imports, sending a message that Europe is no longer willing to blithely accept products with a high carbon footprint from countries that are complacent about decarbonization.[10]

Encouraged by environmentalists and industrialists alike, Europe's Carbon Border Adjustment Mechanism began its transitional phase on October 1, 2023. Initially, importers of goods whose production is carbon-intensive—with a significant risk of leakage (cement, iron and steel, aluminum, fertilizers, electricity and hydrogen)—only need to measure and report the direct and indirect greenhouse-gas emissions embedded in their imports.[11] As of January 1, 2026, however, they will be charged the full market price as set by the weekly EU auctions for tons of CO_2 emitted. That's not a second too soon.

The European Commission explains that "by confirming that a price has been paid for the embedded carbon emissions generated in the production of certain goods imported into the EU, the carbon border adjustment mechanism will ensure the carbon price of imports is equivalent to the carbon price of domestic production, and that the EU's climate objectives are not undermined."[12] Future environmental historians may look back to 2026 as a landmark year for global governance of climate mitigation. But the idea is not entirely new.

It is rarely noted that California has been implementing a border adjustment policy for years. When the California Global Warming Solutions Act was enacted in 2006, almost 18 percent of the state's electricity consumption came from outside its borders. It turned out that more than half of the carbon emissions associated with the out-of-state imports came from power plants with high greenhouse-gas emission intensities.[13] California does not have authority to regulate out-of-state power plants. But its leadership realized that it needed to do something to address the perverse incentives encouraging production of lower-priced, high-carbon electricity outside its jurisdictional boundaries.[14] The rules were amended: when imports come from states that don't have a trading system linked to California's standards, the emissions trading system began to require "first deliverers" of imported electricity to offset carbon emissions from electricity through compliance obligations.[15] This may not entirely solve the "resource shuffling" problem (whereby renewably generated electricity is diverted to a jurisdiction with tougher carbon standards and polluting energy sources are then supplied locally).[16] But it's a start.

Border adjustments have another upside. According to one estimate, such a levy will generate $32 billion annually for Europe's coffers. For countries that are slow to adopt tough greenhouse-gas emissions standards, such policies will soon start to change the calculus at the national and the firm level. Border adjustments may succeed in setting a de facto international price on carbon, giving global carbon markets a major boost.

While the EU, or individual countries, can unilaterally adopt such policies, international agreements offer another way to level the playing field. For instance, the global community could adopt mandatory international standards for minimum energy performance in appliances. Such standards already exist for aviation safety, drugs, and even weapons that cause unnecessary or unjustifiable suffering. There is no reason why climate security should be valued less than personal safety.

The same is true of standardizing electric vehicle charging stations, especially since competing plugs have been developed for vehicles in the United States, Japan, China, the EU . . . and the Tesla universe.[17] Cars and trucks cross borders. The present chaos only aggravates drivers' anxiety about the range their vehicles will be able to drive.[18] Similarly, why not include carbon standards on agricultural imports? Afterall, pesticide standards for food traveling across borders have been around for years.[19]

At the heart of these approaches is the recognition that climate change is a global problem; greenhouse-gas emissions do not adhere to national borders. To wrestle them downward, the international community must work together to decarbonize. This requires a global platform for transboundary coordination and governance. For over thirty years, the UNFCCC evolved as its member nations aspired to meet the myriad challenges presented by climate change.

The Long Road to Paris

In 1997, five years after 154 countries formally first signed the UNFCCC, it was clear that very little had taken place to stabilize atmospheric concentrations of greenhouse gases. Yet the accord was designed to be iterative. Each year since its ratification (aside from 2020, due to the global pandemic), signatories have convened at a "Conference of the Parties," otherwise known as the COP, to review and amend the agreement. A significant two-year effort went into preparations for COP3 in Kyoto, yielding the first protocol approved as part of the UNFCCC process. It was an ambitious agreement—sadly, though, given the politics of the time, too ambitious.

The 1997 Kyoto Protocol contained binding emission-reduction targets for developed countries, which were to come into force by 2005. When aggregated, they were expected to reduce signatory greenhouse-gas emissions in 2012 by 5.2 percent—relative to 1990 levels. With few exceptions, most signatories to the UNFCCC immediately agreed to take on this level of commitment. But those who opted out were important ones: the United States and Canada. At the time, China and India were still considered developing countries, making them exempt from mitigation obligations under the Protocol. Combined, they emitted almost half of the greenhouse gas released into the Earth's atmosphere. The result was one of international environmental law's greatest disappointments.

Nevertheless, one significant contribution of the Kyoto Protocol is that it legitimized international emissions-trading programs, which until then had been limited to domestic air pollution schemes. The Protocol (which for fifteen years was still binding for many countries) allows conscientious parties that *exceed* their emission-reduction targets to sell excess emissions allowances to countries that fail to meet theirs. Besides the EU emissions trading system, numerous national or subnational frameworks for buying and selling carbon have since emerged in Canada, China, Japan, New Zealand, South Korea, Switzerland, and parts of the United States.

The failure to translate the Kyoto Protocol's lofty vision into meaningful global decarbonization is often attributed to the abdication of American leadership. The Republican Party, highly influenced by oil interests, controlled the US Senate, the advice and consent of which were required to ratify the American-brokered agreement. Moreover, many Democrats were also not with the program. Prior to the adoption of the Kyoto Protocol, the Senate passed a resolution introduced by influential West Virginia Democratic senator Robert Byrd by 95–0, rejecting the Protocol if it did not include commitments from developing countries. The Senate wasn't really thinking about the weather anyway. Most of its attention was focused on using White House scandals to bludgeon and impeach a weakened President Bill Clinton. Decarbonization, politically, was a nonstarter. But beyond American political pushback, the Kyoto accord had several other fundamental flaws.

In 2012, Dieter Helm penned a thoughtful autopsy for the protocol in the distinguished journal *Nature*:

> The reasons for the Kyoto Protocol's ineffectiveness are in its architecture. It is based on carbon production, not carbon consumption. It has a mainly European focus. It does nothing to address the immediate problem of global coal burning. It is wide open to free-riding, allowing nations to avoid cutting emissions while others do so, and it has few enforcement mechanisms. These are deep flaws that render the protocol incapable of slowing emissions, let alone reversing them. Fortunately, other, better, bottom-up approaches hold hope for progress.[20]

With the feeble performance of the Kyoto Protocol increasingly apparent, other baby steps were made at subsequent UNFCCC gatherings convened annually around the globe. The bottom line, however, remained unchanged: globally, greenhouse-gas emissions continued to steadily increase. It would take almost two additional decades for the international community to internalize the lessons of Kyoto, accept the limitations of a consensus-based model, and come up with something entirely different. And so it was: with the Obama administration applying the full force of America's global influence, the Paris Agreement was signed at COP21 in 2015.

In order to prevent the worst consequences of global warming, negotiators managed to agree on a ceiling for future temperatures: "limiting global warming to well below 2 degrees Celsius above pre-industrial levels and to pursue efforts to limit the temperature increase to 1.5 degrees Celsius."[21] They also agreed on a different strategy for motivating countries to take action.

In lieu of the international community imposing monolithic emissions reductions, every country was expected "to prepare, communicate, and maintain successive nationally determined contributions (NDCs) that it intends to achieve. Parties shall pursue domestic mitigation measures, with the aim of achieving the objectives of such contributions."[22] In other words, every country needed to figure out what it could do to reduce emissions, how it would do it, and then inform the international community of its contribution to the global effort.

Even before the parties convened in Paris, two years of arm-twisting by US secretary of state John Kerry led to an extraordinary achievement: by the time COP21 opened in 2015 in Paris, prior to the start of the deliberations, 150 countries had already submitted "Intended" National Determined Contributions. In terms of rhetoric, the Paris Agreement was zealous in its support for developing new decarbonizing technologies. Article 10 (1) of the agreement emphasizes one of the rare areas where a true consensus exists among the otherwise highly divided UNFCCC member countries: "Parties share a long-term vision on the importance of fully realizing technology development and transfer in order to improve resilience to climate change and to reduce greenhouse-gas emissions."

Despite the sense of triumph among the diplomats at successfully hammering out a meaningful agreement after years of frustration, the new climate accord engendered no shortage of criticism. By design, the Paris Agreement is not prescriptive. (That way, President Obama could forego Senate approval and thus avoid repeating the Clinton Administration's Kyoto Protocol debacle.) Countries pick their own targets, set their own baseline dates, and take them as seriously (or as insouciantly) as they please. Only reporting is legally binding—and most countries were already doing that.[23] So it was not surprising to discover that when the reduction commitments were totaled up and plugged into climate models, experts found that by the end of the century, average temperatures on the planet were still likely to increase by 3.1°C.[24] Defenders of the new approached explained that, at least now, there was a baseline and point of departure for the planet with specific emission reductions pledged that could be ratcheted down over time.

This voluntary, "bottom-up" approach continues at the time of this writing. Most countries recognize both the limitations of the UNFCCC framework and the space it creates for diverse nations at the negotiating table.[25] For instance, at the 2022 COP27 in Sharm El Sheikh, the UNFCCC recognized the "Loss and Damages" suffered by developing countries due to historic emissions of developed countries.[26] Even if largely symbolic, such decisions contribute to a sense of global solidarity that is necessary to keep many countries engaged.

While the framework can help create consensus, it also engenders intense disagreement between nations. No issue is more consistently disputed than financing. In 2009, parties at the Copenhagen COP15 committed to raising $100 billion per year by 2020 for climate aid to developing countries and for the purpose of "meaningful mitigation actions and transparency on implementation."[27] The amount of funding transferred to the Green Climate Fund to support climate action and deployment of low-carbon technologies steadily increased over the years, but never reached the full $100-billion pledge. When shortfalls between 2013 and 2020 were tallied up, climate funding fell 48 percent below what the international community had promised.[28] Most of these funds are provided as loans anyway. Asian and middle-income countries thus far have been the main beneficiaries, with the poorest nations left behind. Moreover, a 2022 Danish assessment showed that from 2011 to 2018, a mere 6 percent of climate finance provided by rich countries was "new or additional" to existing official assistance for development.[29]

In 2015, a separate, $10-billion fund sponsored by UNFCCC became operational. The Green Climate Fund was created to provide grants to developing countries. For instance, it contributed $154 million to a solar energy project in Zambia that produced two fifty-megawatt power plants; financed upgraded grid infrastructure, photovoltaic systems, and energy storage solutions in the Pacific islands; and supported a climate-resilient water-management project in Nepal, along with many, many more worthy initiatives.[30] It's not a short list. But ultimately, the Green Climate Fund's resources remain woefully inadequate. In 2023, eight years after its establishment, a total of $11.3 billion in pledges and contributions had been received.[31]

In the bottom-up spirit of the agreement, the Paris Agreement extended the $100-billion climate funding commitment until 2025. But contributions remain voluntary. And funds collected under these international agreements are *not* earmarked for developing new technologies and innovation: more than three-quarters are directed to mitigation projects. The other quarter of the funds are earmarked for adaptation efforts in developing and middle-income countries. At the 2021 COP26 in Glasgow, the UNFCCC plenary could urge donor nations to increase their relative support for adaptation but do little more.[32] In other words, the wording of the Paris Agreement may offer a bold vision of accelerating green technology, but it offers nothing in the way of funding climate tech research and development. This, however, does not mean that international climate agreements do not affect innovation.

International Commitments

If international climate accords do not designate money for innovation, how then do they help advance the development and adoption of climate tech? It can be persuasively argued that over the years, the UNFCCC, as well as the more recent NDC process, has created pressure to develop new technologies to meet mitigation commitments. Among the various mechanisms that help countries move from aspiration to implementation, three are particularly important:

- Increasingly stringent emissions-reduction targets;
- Benchmarks for decarbonizing electricity, most notably *renewable portfolio standards*; and
- International cooperation in research and development.

A deeper dive into each approach suggests that however imperfect, international agreements matter.

One central accomplishment of the Paris climate agreement was that countries were expected to declare national emission-reductions targets, along with dates for these goals to be met. Defining an achievable target, along with a concrete plan for reaching it, was the culmination of decades of work, during which the UNFCCC shepherded greenhouse-gas emissions monitoring programs forward. A well-known adage, "If you can't measure it, you can't manage it," is particularly relevant when it comes to greenhouse-gas emissions. By the time the world convened in 2015 for COP21 in Paris, this quantification process was far enough along in most countries to establish discrete emission-reduction goals with a solid, scientific basis.

Six years later, the COP26 meeting was eagerly anticipated. Under the Paris Agreement of 2015, parties were expected to submit new and enhanced NDCs to the climate convention secretariat by 2020. These needed to contain more-ambitious greenhouse-gas emission-reduction goals than the original targets. Just how much more, however, was deliberately vague. By the time of the 2021 Glasgow gathering, 151 of the 193 parties had communicated updated or new NDCs. Within a year, 24 more submitted. The vast majority (94.9 percent) of global emissions were covered under these action programs.

The new round of mitigation pledges showed that the level of ambition for reining in emissions continues to vary dramatically between countries. Many countries are no longer stalling but are already transitioning to a low-carbon economy. The United States, of course, represents the most dramatic turnaround.

In contrast to President Donald Trump's dismissive attitude toward global mitigation efforts, at the 2021 Glasgow COP26, the Biden Administration pledged that by 2030 American carbon emissions would be 52 percent lower than in 2005. This was roughly twice the commitment that President Obama was able to make only six years earlier. The EU went a step further, pledging a 55 percent reduction over the same time frame.[33] And in 2020, Denmark's parliament passed a law to reduce climate emissions 70 percent from 1990 levels by 2030.[34]

Not every country has been so forthcoming with its commitments. Responsible for 27 percent of the global greenhouse-gas emissions, China is the world's largest contributor to carbon in the atmosphere. A significant amount (14.6 percent) of those emissions is driven by consumer goods that are exported for foreign consumption; if China's manufacturing hubs constituted a separate country, their CO_2 emissions in 2012 would rank fifth in the world.[35] Nonetheless, the remaining 23 percent of global emissions associated with meeting domestic Chinese needs is still a huge carbon footprint—60 percent higher than the United States, the second-largest emitter. Overall, emissions in China are expected to continue to increase beyond 2030. While the Chinese have promised national "carbon neutrality" by the year 2060, China's leaders' unwillingness to announce what year the country's emissions would peak and begin to fall belies the seriousness of their pledges.[36]

When the projected emissions of each nation were tallied up after the new 2021 "Glasgow" NDC commitments, relative to a baseline of 2010, by the year 2030, global emissions were still set to increase by 10.6 percent. This constitutes a profound disappointment for many, not least the United Nations climate team.[37] After all, there is no guarantee that countries will actually meet their targets, let alone exceed them. Leading up to the 2022 COP27, the UN tried to offer a positive spin on this dour news, reporting that the anticipated increase was a full 22 percent lower than the 13.7 percent rise that was projected just a year earlier.[38] This represents a modest improvement. But clearly it is not good enough.

With such anemic progress, should the world give up on international climate accords? Of course not. Without the UNFCCC's global platform, it is unlikely that any of new targets would have been declared. More importantly, these goals have pushed nations around the planet to begin the process of decarbonization. Proof of the seriousness of America's new commitments can be found in the 2022 Inflation Reduction Act. Its $369 billion for funding energy security and climate-change programs over the next decade is unprecedented. The American government, it seems, is taking international commitments seriously.[39]

The United States is hardly alone. In order for Europe to meet its 55 percent emissions-reduction target and ultimately "become the first carbon-neutral continent on the planet," the EU formally adopted a Green New Deal. This comprehensive strategy promises to transform transportation, manufacturing, electricity, and food systems across the continent,[40] with €600 billion allocated over the next seven years to fund a complete societal makeover.[41] Similarly, by its own admission, Denmark became such a passionate, early decarbonization adopter because of UNFCCC dynamics. After hosting the 2009 COP15 in Copenhagen, it began working on a bold national strategy to reduce emissions from agriculture, industry, and even shipping. By 2020, 80 percent of the power pulsing through the country's electrical sockets came from renewable sources.[42]

The story is the same all over the world. When national leaders realize that they will have to come and make a speech about climate in front of their peers at a United Nations event, even the reluctant can make surprising commitments. Israel, long a laggard in the area of climate mitigation policy, is a good example. Literally on the eve of its delegation's departure to the Glasgow COP in 2021, the government upgraded its emissions-reduction target, pledging to join the growing community of nations committed to becoming carbon-neutral by 2050.[43] Naftali Bennett, Israel's prime minister, candidly acknowledged the role of peer pressure in the decision: "We had to do it sooner or later. So it might as well be sooner."[44]

International accords have also encouraged developing countries to engage in climate mitigation. In 2015, numerous African nations for the first time made commitments to reduce emissions.[45] This process continued at Glasgow with new pledges for decarbonization received from countries like Mozambique,[46] Mauritius,[47] and even "least developed countries" like Ethiopia and The Gambia.[48] South Africa pledged an interim emissions goal 32 percent lower than its NDC in Paris.[49] Bangladesh promised to reduce its total emissions by 6.5 percent by 2030, anticipating that 95 percent of this reduction would take place in the energy sector.[50] If these pledges seem less dramatic than those of the climate champions, perhaps that's because the developing countries' baseline emissions per capita are already so low. Even so, a ratcheting-down process has begun.

One striking phenomenon produced by the Paris Agreement is the high number of countries willing to declare that in the distant future they would become carbon-neutral. By 2023, some 140 countries had committed to a net-zero emissions target by 2050, with China making the pledge for the year 2060.[51] "Net-zero" implies a commitment to balance the total amount of greenhouse gases emitted with the amount removed. In practice, this will have to be achieved by an extremely broad range of decarbonization measures—starting with renewable electrification—and supplemented by removal of any remaining carbon emissions from the atmosphere. The net-zero pledges, in fact, represent a dramatic diplomatic achievement. By 2023, countries that together account for more than 90 percent of global GDP and 88 percent of global GHG emissions publicly promised to eliminate all emissions by the mid-twenty-first century.[52] Suddenly, the direction in which the world economy will be moving during the coming quarter century became clear. And it appeared to be far more sustainable.

It is far too early for congratulations. Net-zero commitments have correctly been defined as aspirational in nature. But they are not "nothing." Once in place, they need to be accompanied by a series of incremental benchmarks to get there. For instance, most models assume that *globally*, emissions from the power sector need to fall by 7 percent a year by 2030 if the planet is to stay on track to reach net zero by mid-century.[53] This should become a prescribed part of the next NDC submission process: quantitative commitments for significant future reductions, creating a tangible roadmap to net-zero targets.[54]

Renewable Portfolio Standards—The Potential of Quotas

Over the coming decade, perhaps the easiest and most reliable way to achieve national greenhouse-gas emissions-reduction targets is a rapid transition to renewable energy. The burning of fossil fuels to generate electricity is responsible for over 25 percent of greenhouse-gas production globally: 19 percent comes from coal; 6 percent from natural gas; and 1 percent from oil.[55] Eliminating these emissions expeditiously has emerged as the most popular mitigation strategy in national action plans, followed by improving energy efficiency. Of the 144 NDC plans resubmitted in the 2021 cycle, 109 focused on increased electricity generation by renewables. That said, only 23 countries were willing to commit to a renewable capacity higher than 60 percent of their electricity supply.

In the United States, for many years "renewable portfolio standards" (RPS) programs have constituted an effective way to lower greenhouse gasses from electricity, which currently account for 25–30 percent of the nation's total emissions. Portfolio standards require that a certain percentage of a state or country's electricity generation comes from renewable sources, including wind, solar, hydroelectric, and geothermal power. Often the standard is imposed directly on an electric utility, stipulating that it must derive a set percentage of its power from renewable sources. With thirty-one states and the District of Columbia adopting them, RPS became a way for states to take action even when the national government lacked a coherent climate strategy.[56]

The first renewable portfolio standard goes back to 1983. Contrary to prevailing stereotypes that assume that liberal urban dwellers support renewable policies while conservative rural citizens oppose them, diverse communities across the United States have embraced clean electricity. Iowa was the unlikely first US state to impose a renewable quota on its utilities. Its moderate Republican governor realized the potential of wind farms for creating an additional income stream for economically pinched farmers in a state where 20 percent of the workforce is associated with agriculture.[57]

Not surprisingly, there was fierce initial opposition from power companies to early RPS proposals. They claimed that forcing them to purchase electricity from more-expensive renewable sources would drive up consumer prices. Such resistance is often finessed by politicians who simply set initial standards at a modest level that doesn't challenge grid capacity or corporate profits (e.g., mandatory renewables for only 1 percent of electrical generation). But it did not take long for RPS policies to enjoy public support, with mandatory renewable percentages steadily raised because of their environmental benefits and economic potential.

Portfolio standards were soon adopted across the United States and then in dozens of countries, from China, India, and Japan to Brazil, Australia, and the United Kingdom. They typically began modestly with low renewable requirements, and then gradually climbed as the price of renewables dropped and the urgency of climate mitigation increased. As part of its strategy to wean itself from Russian oil and gas by 2027, the EU raised its 2030 renewable goals from 32 to 42.5 percent with many countries going further.[58] Among the first environmental policies that President Joe Biden adopted after taking office was an 80 percent renewable energy target by 2030, with the expectation that the United States would reach carbon- (and pollution-) free electricity by 2035.[59] The island of Hawaii was already way ahead: in 2015 it became the first American state to adopt a 100 percent RPS target.[60] Kenya has set a goal of having a 100 percent renewable electricity system by 2030 and will probably get there earlier, as it already is 80 percent of the way there.[61] Renewable portfolio standards are also credited with driving investment and innovation in renewable energy technology.

In practice, RPS requirements can take different forms. They often appear as "production targets," set in megawatt-hours (MWh). This is essentially a *performance standard*.[62] In most jurisdictions, to meet the standard, electricity providers can either produce renewable energy themselves or acquire renewable energy certificates from other renewable energy generators. (If an electricity provider is required to source 10 percent of its electricity from renewable sources, it can either generate 10 MWh of renewable energy or purchase 10 corresponding certificates.) The primary advantage of adopting a more flexible regulatory format is incentivizing producers to maximize renewable electricity production. The alternative is simply to set a megawatt (MW) standard for renewable facilities, based on their potential capacity to generate energy. This may be simpler for regulators to oversee, but it does not ensure that this amount of electricity will actually be generated. Electricity producers might find it more convenient to site a renewable facility in an inexpensive but less than optimal location for absorbing sunlight or capturing wind.

When governments set a renewable portfolio standard, the rule generally includes a list of approved technologies. This can be a source of controversy. For example, should nuclear-powered electricity be accepted as meeting an RPS? To meet a quota, can electricity be imported from an area where renewable generation is cheaper, due to more-advantageous climatic conditions? (If reducing carbon emissions is a priority, then the answer to both is "yes.")

Like any nonvoluntary intervention, there need to be consequences for noncompliance should utilities choose to take a path of least resistance. Policy makers are encouraged to think carefully and select the right level of enforcement for their RPS programs, lest they come off as too lenient and encourage noncompliance—or too draconian and endanger public support. In a manual produced by the US Department of Energy, the federal government recommends that "a penalty or other means to address enforcement of the requirement can help ensure future annual targets will be met. . . . Without these measures, critical investments may not be made due to lack of investor confidence, and the RPS would function more as an aspiration rather than a solid objective on which renewable developers and investors could rely."[63]

Penalties for noncompliance can take many forms, depending on the level of severity that legislators—or government agencies—are willing to impose. Alternative compliance payments are common. These are fines requiring a noncomplying utility to pay a sum (e.g., $500) for every MWh of renewable electricity they are short in producing. Penalties can differ by the type of renewables or be based on the profits which noncomplying companies accrue by dodging the requirement.

While renewable performance standards typically are considered one of the most effective policies to cut emissions, it is important to step back and measure just how well they work. Researchers have come to widely different conclusions. One study from University of Chicago researchers, appropriately titled "Do Renewable Portfolio Standards Deliver Cost-Effective Carbon Abatement?," concludes that they do . . . but only modestly. After comparing twenty-nine states that adopted RPS requirements with twenty-one that did not, the team reported that twelve years after adoption, on average, there was a 5 percent increase in renewables' share of electricity supply in states with renewable portfolio standards. They also found a 2.0¢ per kilowatt-hour (kWh) or 17 percent increase in the price of electricity.[64] A 2016 study by Lawrence Berkeley National Laboratory found an annual average of 5,600 megawatts (MW) in new renewable capacity due to renewable portfolio standards. (The researchers also noted that compliance with RPS standards led to nearly 200,000 US jobs in 2013 and drove more than $20 billion in gross domestic product.)[65]

Another study, however, was significantly less encouraging. Michigan-based researchers looked at twenty years of RPS experience in twenty-eight American states, asking whether the standards might actually stunt renewable electricity development, with utilities choosing not to go beyond mandatory targets. Their answer turned out to be yes—unless a state is particularly fortunate in enjoying excellent natural conditions for generating renewable energy. Nonetheless, the study points out the obvious: the standards offer a modicum of certainty that renewables will be part of the mix.[66] After decades of stagnation, this was important for creating some momentum for a decarbonized grid.

Still another comparative RPS evaluation from Hood College in 2021 was more unequivocally positive, crediting the standards with about a third of the increase of American renewable electricity capacity. Results also suggest that not all renewables are created equal; significant increases for solar and wind were measured due to RPS programs, contrasting with biomass and geothermal, where the effect has either been insignificant or even negative. Size also matters: the larger the area covered by an RPS policy, the greater the increase in renewable adoption.[67]

While these studies clearly show that portfolio standard programs have room for improvement, they also confirm RPS's important role in decarbonization, especially when utilities need a push. To return to this chapter's overall theme, renewable portfolio standards have been spurred on by international climate agreements. Researchers cite many reasons why renewable portfolio standards are adopted: voter ideology (or government officials' inclinations), air quality, renewable energy potential, per capita income, and the price of electricity. But national commitments made pursuant to international agreements emerge as a key factor.

The International Renewable Energy Agency reviewed all the original climate reduction plans submitted at Paris. It found that 71 percent of the 190 plans included promises to reach clear percentage targets for installation of renewable energy. This is roughly three times more than those with quantifiable commitments for emission reductions from transportation or heat. The agency calculated that if the 134 countries who made these pledges implement them, an additional 1,041 gigawatts (GW) of renewable electricity would be generated by 2030, increasing clean power production worldwide by 42 percent.[68] Mitigation pledges to the international community make a difference.

When countries consider how to meet emissions targets, it turns out that renewable energy is among the most cost-effective things they can do. That's why renewables are expected to account for 90 percent of global electricity capacity expansion during the next five years. Politicians more concerned about employment can point to the fact that the standards help create jobs—to the tune of 600,000 jobs/year, with 12.7 million people employed globally in 2022.[69]

Climate Cooperation

In addition to putting pressure on nations to innovate, international climate agreements have a key role to play in finding the right balance between competition and cooperation in technology research. Clearly, competition is important. It motivates scientific teams from different countries and institutions through financial rewards and prestige. At the same time, cooperation helps build knowledge by expanding the pool of experts, prioritizing funds for the most promising solutions, and sending a healthy signal to investors about potential gains from greater economies of scale. When countries become too focused on beating other countries technologically, they risk wasting scarce capital, human resources, and credibility. Duplication, at some point, becomes inevitable.[70]

It is usually easier for politicians to get behind the "competitive" side of this continuum. After elected officials or bureaucrats mobilize significant resources for a major demonstration project or research initiative, they may understandably be loath to share the details of important breakthroughs in order to maximize the payback on their country's investment (or their own political effort). A goal in international climate negotiations is to ensure that competition helps drive the decarbonization transition and not serve as a brake.

The history of science shows the enormous value of research collaboration. Indeed, the much-heralded recent breakthrough in the area of nuclear fusion research was only possible because two very different laboratories on opposite sides of the United States worked together: Rochester's Laboratory for Laser Energetics brought its laser-driven implosion techniques to the Lawrence Livermore National Ignition Facility's nuclear team.[71] It took 192 laser beams to deliver two million joules of ultraviolent energy to a tiny fuel pellet and create fusion ignition on December 5, 2022. The results offered a rare moment of hope that today's investment in basic scientific research may actually allow future generations to enjoy abundant, emissions-free electricity.

Because research is increasingly interdisciplinary, cooperation needs to extend beyond different labs to different scientific fields. Complex problems invariably require broad consortia. One example is the EU's Horizon program, which is based on cooperation by universities—with 35 percent of total funding earmarked for climate research. As an incentive to find appropriate partners, the program offers a massive budget: €95.5 billion between 2021 and 2027.[72]

UNFCCC meetings and COPs provide a global venue for bringing such experts together. In fact, the COPs often have the feel of a matchmaking forum, with researchers and entrepreneurs conducting the equivalent of a speed-dating exercise as they mingle and exchange their ideas with hundreds of attendees. This is an informal process that could easily be formalized through better organized meetings of specific emission sectors. The existence of Zoom and other remote conferencing programs means that it could happen without the mass attendance at the COPs, with their colossal carbon footprints.[73] Whether virtual or in-person, such international gatherings offer an opportunity for collaboration that is badly needed for advancing technical solutions.

Nonbinding international cooperative frameworks are already common in scientific research and are a favorite talking point for foreign ministers seeking positive media coverage when meeting their counterparts on foreign junkets. To avoid replication and to maximize limited resources, it makes sense to establish more large, multilateral, commercial-scale pilot projects or testing sites and increase their geographical distribution.[74] International cooperation in research and development already takes the form of research teams sharing multinational testing facilities.

Over the years, world leaders came to realize that the formal UNFCCC process is not the only avenue to pursue cooperation. Indeed, as the UN's climate agreement became more politicized and progress became agonizingly slow, countries of like mind began to organize climate initiatives that circumvent the formal UNFCCC forums and enlist the private sector. Ironically, the annual COPs provide an ideal launching ground.

Cooperative initiatives can also establish systems to deliver power across borders, making clean electricity more widely available and reducing costs dramatically. At the Sharm El Sheikh COP27, a memorandum of understanding (MOU) between Israel, Jordan, and the United Arab Emirates did just that: Israel agreed to transfer copious quantities of its low-cost desalinated water to Jordan in return for solar-generated Jordanian electricity.[75] Creating regional networks of hydrogen pipelines might also send an important signal to companies as this energy source moves forward. In general, "gigafactories," which employ economies of scale, can dramatically reduce the costs of technologies like batteries or wind turbines.

At the Glasgow climate conference, three significant initiatives debuted with great fanfare that promised to accelerate progress in climate tech: the Breakthrough Agenda, Mission Innovation 2.0, and the First Movers Coalition.[76] None are binding, yet each appears to be generating more concrete results than the numbing debates over the language of UNFCCC resolutions at the COP plenary.

Mission Innovation was actually launched originally in 2015 when US president Barack Obama, French president François Hollande, and Indian prime minister Narendra Modi joined Microsoft founder Bill Gates as he announced his Breakthrough Energy coalition. Rather than take a chance on sluggish, multilateral negotiations, twenty countries simply announced that they would double their respective budgets for clean energy research and development. Collectively, the Mission Innovation member countries are responsible for 75% percent of global CO_2 emissions from electricity and 80 percent of global clean-energy R&D investments.[77] Since the announcement, their climate research budgets definitely increased—but not nearly enough. One assessment suggests that the initial financial commitments made to fund Mission Innovation fell $50 billion short of the collective goal.[78] International inertia helped to fix this.

Five and a half years later, prior to COP27, the coalition expanded to include twenty-five countries. Its members are now responsible for 90 percent of all public investments in clean-energy innovations. Strategically, Mission Innovation seeks to support research and development that reduces the costs of low-carbon technologies in order to reach a "tipping point" where widespread adoption makes economic sense. Recognizing the many technological gaps currently preventing the transition to "net-zero emission" economies, partner countries pledged to increase research funding by an additional $5.8 billion. During the 2021 launch, Mission Innovation also promised to draw on collective expertise to make advances in zero-emission shipping, clean hydrogen, carbon monoxide removal, net-zero industries, urban transitions, and integrated biorefineries.[79]

Another initiative to emerge from the 2021 Glasgow COP26 was the First Movers Coalition (discussed in length in chapter 7). The coalition is a public–private partnership between the United States and the World Economic Forum to elicit support from the world's leading corporations to buy low-carbon technologies that are still emerging. The primary objective is to harness the purchasing power of the private sector to decarbonize seven "hard-to-abate" industries: aluminum, aviation, chemicals, concrete, shipping, steel, and trucking—along with carbon-removal technologies. Together these industries make up 30 percent of total greenhouse-gas emissions.[80]

The final example of an initiative that is linked to, but not a formal part of, the UNFCCC process is the Breakthrough Agenda. The initiative also debuted at the Glasgow COP, endorsed by forty-five world leaders whose countries produce more than 70 percent of the planet's gross national product.[81] Essentially, the Breakthrough Agenda is an ambitious plan for countries, civil society, and research institutions to cooperate on issues related to climate, prioritizing research and development.[82]

Once again, rather than create a formal UNFCCC agency that would quickly be entangled in North–South bickering, the Breakthrough Agenda is an independent voluntary organization. And rather than specifying emissions reductions, it focuses on making clean power the most affordable and reliable option for *all* countries to meet their power needs efficiently by 2030. It calls for "zero-emission road transport" to become the new normal—in *all* regions by 2030. It seeks to make zero-emissions steel the preferred choice for global markets and affordable hydrogen available globally, as well as making climate-resilient agriculture the most attractive option for farmers worldwide.[83]

These essentially extralegal declarations are not enforceable. But the history of the UNFCCC suggests that for the foreseeable future, global climate commitments will remain voluntary anyway. That doesn't mean they don't matter. They send a message. They also open the door for more-prescriptive, top-down provisions in future rounds of UNFCCC negotiations. Perhaps most importantly, they create confidence in future markets for climate tech products. (This will be particularly important for creating large enough markets for the deployment and trade in hydrogen.)[84] Using the stature and the excitement of the UNFCCC gatherings to inspire (and pressure) partners from the corporate world to help create this demand constitutes a meaningful, but often unappreciated, contribution to global efforts.

Making climate tech work

Policies that Drive Innovation

Very Good

bc. kuldeep

I star 1

2025-10-14 20:12:24 (UTC)

Zoom Books Company

Aisle 31 Bay 1 Shelf 5

ZBM1V1H

ZBV 1642833386X VG

120

3969451

0.70 in

1,588,825

ZBV 1642833386X VG

digital order: 3069M 451

XXXXX

In short, the benefits of international climate agreements go beyond the details of their formal provisions. Alongside debates over global targets, as well as the ineluctable politics that color UNFCCC gatherings, the United Nations' annual climate convention creates a platform where cooperation and goodwill are often the norm in the world's common effort to combat the climate crisis.

It is not very hard to point out the many shortcomings of the UNFCCC. At the end of the 2023 COP28 in Dubai, the compromise decision only called for "transitioning away from fossil fuels in energy systems . . . in this critical decade" rather than a clear phase-out.[85] Low-lying island nations were deeply disappointed by the "litany of loopholes" that the consensus agreement allowed.[86] Yet for the first time, the international community formally put the oil and coal industries on notice that their old business model was on its way out. The decision, supported by nearly 200 countries, also explicitly states that limiting global warming to 1.5 °C requires sustained and rapid reductions in global greenhouse gas emissions of 43 percent by 2030 and 60 percent by 2035 (relative to the 2019 levels) in order to reach net zero carbon dioxide emissions by 2050.[87]

In the absence of a global climate accord and robust international cooperation, the transition to net zero would be significantly slower and more expensive. National leaders would face even fewer expectations and less peer pressure. Supply chains for critical technologies like batteries could more easily be interrupted. Given the high start-up costs for climate tech demonstration projects and testing, without collaborative efforts, innovation would suffer. Climate technology innovation is essentially a race against time. The tailwinds provided by a UN climate convention, however imperfect, should not be underestimated.

CHAPTER 3

Jumpstarting Research and Development

When societies face problems that require technological solutions, it makes sense for public resources to support research to develop them. The outcomes of research programs, however, are inherently uncertain. Experiments are not always successful; hypotheses are not always correct. By definition, research takes scientists into the realm of the unknown. And even when everything works, funding is received, scientists recruited, results published—it still takes years before marketable products, processes, and practices are ready to go. Uncertainty is actually built into the process. It is often argued that the greater the certainty of an R&D initiative's results, the more modest its impact is likely to be—making any innovation incremental at best.

For transformative progress, researchers must embrace ambitious research projects whose results will not always be satisfying. The heartbreaks and elation of research and development offer the only path forward if the world is to develop the full range of decarbonizing technologies needed to stabilize greenhouse gases.

It is also worth noting that the UN has come to add a stage in the climate tech research process that it now calls "RD&D." After *Research and Development, Demonstration* is deemed essential to show "the use of the product in actual field conditions where its performance and feasibility can be demonstrated and evaluated by actual or potential users."[1] Inventions, it seems, are not enough. Decarbonization requires that emissions-intensive products and processes be expeditiously replaced by low- and zero-carbon alternatives. For this to happen, there must be a marketing and dissemination strategy that shows their advantages.

What can be done to ensure that the right level of resources is available for research by the right experts, on the right topics? There are two schools of thought on this matter. The first is that government should *push* technology forward by investing directly in research, while the opposing perspective is that government should use regulations and incentives to *pull* technology along: in other words, create the demand for a new technology that private companies will then use their own resources to invent.[2]

Proponents of the *"demand-pull"* approach contend that the private sector is better than the government at understanding market demands and developing new technologies to meet them. Yet history shows that relying solely on market forces does not always produce the best results for society. Take, for example, the long and languid record of photovoltaic electricity development, which for years lagged because private companies were inadequately funded. It is only natural for corporate executives to hesitate to invest in research with high costs and uncertain outcomes when safer and more lucrative opportunities exist.

Even when firms are interested in conducting research and developing an important technology, they frequently "take a pass." There is no shortage of reasons: short investment horizons, pressures to generate immediate profits, and market uncertainties, along with concern about "knowledge spillovers" that threaten a full return on their intellectual property investment. All of these militate toward an outcome that will be "too little, too late."

The truth is, if climate innovation is to race forward, both approaches are needed. This chapter focuses on the first: *"technology-push"* policy instruments that aim to reduce the cost and accelerate the pace of climate tech research and development. These interventions encourage companies to tackle the uncertainties of creating socially beneficial technologies through programs such as research grants, low-interest loans, tax exemptions for firms, and the establishment of government laboratories and incubators.

Historically, governments tend to prefer the indirect route to technology development over doing it themselves. Over the years, they have placed their faith in a "pipeline" model of innovation: funding basic research at universities and research institutions on the assumption that eventually these discoveries will give rise to valuable new products, processes, industries, and jobs. There are innumerable success stories that validate this cautious, peer-reviewed, multistage approach. A few such stories are included in this chapter as examples.

There are also times of crisis when an emergency requires a more focused, urgent, and substantial research effort. The World War II Manhattan Project's atomic bomb immediately comes to mind; the race to a vaccination during the Covid pandemic was one such moment; the climate crisis should be seen as another. A slow, incremental strategy simply will not generate the revolutionary technologies required to reach net-zero emissions in time to avert a climate catastrophe. Given the critical nature of the mitigation challenge, climate tech research will need to be different: governments must be far more prescriptive and take more chances in directly backing innovative projects.

There are, of course, risks to abandoning the standard precautions typically required by replicable scientific studies (and peer review). It is important to learn from the many examples of investment in well-intentioned research that failed. The American government's investment in the California solar-panel startup Solyndra (described in chapter 1) is by no means unique.

Norway's ill-fated efforts at carbon capture may be the most spectacular fiasco involving public funds and climate technology to date. In 2007, Norway's prime minister, Jens Stolenberg, waxed visionary about a project to capture emissions from the Mongstad oil refinery, calling it a grand project for the country. "This will be our moon landing," he declared. Six years later, when the cost estimates for the project had quadrupled, crossing the billion-dollar mark, it was canceled. The investment certainly spawned several academic publications[3]— but, for now, precious little carbon sequestration.[4]

It would be wrong, however, to see these examples as proof positive that government intervention is ineffective. Public support of research and innovation in the private sector has also led to spectacular successes. For instance, Denmark illustrates that with the right policies, a society can "punch above its weight" and produce impressive results. Similarly, the case of ReSource Chemical Corp. in Oakland, California, shows both the impact of present government R&D programs and the constraints of existing policies in helping climate tech startups overcome their challenges. (Both cases and their implications will be described.)

Moreover, had the Department of Energy not disregarded Tesla Motor's substantial previous losses and provided it with an audacious $465-million loan guarantee in 2009, Tesla's Model S would never have made it out of the production lines to become *Motor Trends* 2013 Car of the Year. Tesla paid back the loan nine years early. The leap of faith didn't cost US taxpayers a penny, and it transformed the EV industry. The reasons to take risks and support innovation defy a narrow cost–benefit equation. As Princeton professor Dani Rodrik explains:

A promising new technology may be worth supporting even if it does not generate many jobs; employment objectives are better served through other policies. And it may be worth supporting even if the pioneering investor ends up bankrupt; if the technological learning and spillovers from the pioneer spawn a new industry, its own commercial failure is of little consequence.[5]

One thing is certain: the invisible hand of the market will not steer the world economy to a net-zero future. This will require many, many applied research endeavors that address myriad mitigation challenges. They will need to be choreographed and coordinated. This chapter considers public policies that attempt to do just that by accelerating climate tech through research and development.[6]

Growth in Funding for Climate Research

There is little doubt that climate technology research is a growth industry.[7] In recent years, governments around the world have ponied up tremendous sums to support basic and applied research to develop decarbonizing technologies. In 2009, for example, South Korea announced plans to allocate $85 billion for clean energy (a full 2 percent of GDP) to advance the country's "green transition"; this included a significant investment in technological research.[8] Applied R&D was a key part of British prime minister Boris Johnson's £12-billion "green industrial revolution."[9] The Australian "Technology Investment Roadmap" is to be funded at AU$20 billion by 2030. It prioritizes development of low-emission technology.[10] In 2022, Israel's cabinet created an $870-million fund for supporting climate technology.[11] China's dominance in PV manufacturing is largely due to a substantial government-financed research program in large-scale photovoltaics.[12] Germany's decarbonization research agenda has always been extensive and includes wind, biomass, geothermal, storage, refrigeration, and grid technologies. The EU is set to "reinject" billions of Euros accrued annually as part of its Emissions Trading System back into industry to help finance climate innovation.[13] Some 35 percent of funds in the EU's enormous Horizon Europe initiative is to be earmarked for climate-related research.

With the passage of ambitious statutes and the allocation of hundreds of billions of dollars in funding, the Biden administration has raised the proverbial bar, dramatically upping America's climate tech R&D game.[14] Few historians today can name the specific laws which actually allowed Franklin Roosevelt to lift the United States out of the economic morass of the Great Depression.

But everyone knows that together they created a "New Deal" for the country. Similarly, if humanity succeeds in decarbonizing the global economy, it is unlikely that future historians will remember specific climate legislation passed under the administration of Joe Biden. Nevertheless, it should be noted that President Biden picked up the gauntlet, spearheading a trio of laws during his first two years in office to accelerate climate tech through domestic American manufacturing.[15]

While much is made of the $369 billion allocated for climate-related activities under the Inflation Reduction Act, many experts believe that the Infrastructure Investment and Jobs Act (IIJA), often known as the "bipartisan infrastructure bill," may be even more important for stimulating climate tech R&D.[16] That legislation brings together some $550 billion in federal investments in roads and bridges, water infrastructure, resilience, the Internet, and more.[17] It also contains numerous provisions related to research, with a focus on renewable energy and decarbonization. The dizzying list of programs in the Act follows:

- $21 billion to fund research and develop clean energy technologies, including hydrogen, advanced nuclear, and electric vehicles;
- $9.5 billion to support R&D focusing on energy-storage technologies, including batteries and other forms of energy storage;
- $3 billion in R&D funding to advance low-carbon hydrogen technologies;
- $2.5 billion in R&D funding for carbon capture and storage (CCS) technologies; and
- $1.5 billion in R&D funding for advanced renewable energy technologies, including geothermal, marine, and hydrokinetic energy.

These massive allocations point to a broader worldwide trend of growing financial investment in addressing climate change. By 2020, aggregate resources had reached an estimated $632 billion a year. A detailed assessment by the Climate Policy Initiative, a leading think tank in the field of green economics, calculates that the public sector accounts for 51 percent (US$321 billion) of climate finance. Because of the lack of reliable data, identifying the percentage of public funding directed to climate technology research on a global scale turns out to be complicated. Nonetheless, an International Energy Agency report states that combined R&D budgets for renewable energy, energy efficiency, and carbon capture and storage (CCS) technologies have steadily increased, totaling over $20 billion in 2021, even before the well-funded American legislative mandates.

Given the inherent riskiness and delayed returns in climate technology research, it is safe to assume that most of these funds came from government programs. But even this may not be enough. The International Energy Agency's net-zero scenario calls for mobilization of $90 billion to complete a broad portfolio of demonstration projects by 2030. As of 2021, only $25 billion was budgeted for the coming years.[18]

Government's Delicate Dance in Sponsoring Climate Research

It is not just the amount of money being thrown at the problem that will determine whether the technologies critical to decarbonizing the economy will become available. How the money is spent is no less important. Government policies to promote research in climate technologies generally fall into several categories. There are governmental programs to support research in academic, public, or private research initiatives; there are those that fund incubators/accelerators that specialize in developing decarbonizing technologies; there are still others who create research partnerships between the climate tech sector, government, and occasionally philanthropy; some help climate tech startups and entrepreneurs through competitive grant programs; while others provide financial incentives for corporate or entrepreneurial R&D via tax credits, research contracts, loan guarantees, and direct subsidies.[19]

Entrusted with public funds to support research, officials are challenged to find a balance between their responsibility to taxpayers and the need to give researchers latitude to work creatively. Professor Mano Trachtenberg is an international expert in innovation policy who also played a key role in advancing Israel's economic progress while heading the country's National Economic Council. In that capacity, he advocated strongly for keeping the government's innovation program technologically "agnostic." For instance, in 2008 he opposed substantial government funding to launch Better Place, a local electric vehicle venture that sputtered out after much initial excitement. With an academic's appreciation for history, he explains how hard it is to predict technological trends: "There was a fleet of 3,000 electric taxis in New York around the turn of the twentieth century. It would have been very unwise to bet on the internal combustion engine at that point in history." In retrospect, the emergence of Tesla and its superior-range batteries validated his skepticism about backing Better Place.[20] Such a gamble by the Israeli government would have been a bad idea—a classic case of "first is worst."

Yet being entirely agnostic is foolish when there are specific technological gaps that experts believe must be filled. After analyzing the state of the science and the barriers to a sustainable path forward, governments have an obligation to define their priorities for advancing climate mitigation. For instance, even with the drop in the price of renewables, conserving electricity generally costs less than producing it. It makes perfect sense for governments to prioritize research to improve the efficiency of heating, ventilation, and air-conditioning (HVAC) systems, which consume over 50 percent of the electricity used in homes, over R&D designed to reduce the 2 percent used for cooking.[21]

Anders Eldrup, Danish innovation czar, concurs with this view. After a long career in government as head of the Danish finance ministry and then as CEO of Denmark's leading wind energy corporation, Eldrup took over Innovation Fund, Denmark. The fund is tasked with catalyzing the country's "lean and mean" climate tech ecosystem. He summarizes this view succinctly: "A sound strategy is for governments to prioritize certain technologies, but not necessarily the recipients of government support."[22] Winning *solutions* should be selected, rather than winning *firms*.

This approach seems to be the trend with the most recent climate legislation. For instance, the US Energy Act of 2020 establishes a new program called the Carbon Utilization and Storage System (CUSS) program, which provides funding for research and development of technologies to capture, use, and store carbon dioxide. But it doesn't specify how and to whom the funds will be allocated.

In short, a delicate balance must be found between *oversight* and *control*. Political scientists Allan Bentley and Joanna Lewis wisely argue that government agencies need to be embedded in, but not "in bed" with, business: "The right model lies between arm's-length and capture."[23]

To attain that balance, the nature of the partnership must be defined. There is a legitimate debate as to whether governments should demand equity when they assist startups at early-stage research. In the case of Tesla, for instance, critics claim that taxpayers deserved a fair share of the copious profits that Tesla reaped after the American government guaranteed the company's loans.[24] Others take a jaundiced view of the government's siphoning off dividends from successful research because it stymies innovation.[25] Different philosophies lead to different national policies.

Israel's ability to galvanize innovation gained it the epithet of "Start-up Nation."[26] Government policies helped. During the past decade, the government's Innovation Authority has invested in some 920 companies, providing financing for approximately 1,500 innovative research projects.[27] The Authority decided early on that the social benefits, jobs, and tax revenue from these projects were a sufficient return for taxpayers, and did not require dividends from the companies. Indeed, the Authority *wants* scientists and engineers to enjoy windfall profits when they succeed so that entrepreneurs will continue to be highly motivated.[28]

In recent years, the Israel Innovation Authority has expanded its support for climate tech startups, but this has little to do with saving the planet and everything to do with creating business opportunities for the country. It is interesting to note that Denmark, after many years of taking the Israeli approach, has recently begun taking equity in companies that the government sponsored in their early stages of development. This question of whether or not government should have an equity stake in private ventures is particularly germane for companies that spend time in a government-sponsored incubator or accelerator.

Incubators and Accelerators

While it is not unusual for the terms *incubator* and *accelerator* to be used interchangeably, there are real differences between them. *Incubators* offer a home to early-stage startups,[29] and are designed to nurture an incipient firm by providing scientific counsel, management training, legal advice, peer networking, and investor access.[30]

Accelerators, by contrast, support startups at a later stage in their development, when companies can already produce a viable product and articulate a clear market focus. Life in accelerators tends to be intense, driven by a goal of catapulting emerging startups to the next level. The most promising technologies are selected, and then mentors, laboratories, and connections are provided so that climate tech entrepreneurs can enjoy a safe space in which to develop a profitable business.

Neither incubators nor accelerators are free. Generally, incubators are supported through government grants, corporations, venture capital firms, or universities. The same is true for accelerators, although a higher percentage are sponsored by venture capital groups, which frequently provide seed funding in exchange for equity. It often proves to be a lucrative association.

The list of companies emerging with fully formed, innovative technologies from incubators and accelerators reads like a pantheon of climate tech superstars. One example is Sunrun, which received a modest $1.6 million grant to launch its automated PV installation project as part of the US Department of Energy's Sunshot incubator program in 2013.[31] The company went public in 2015, soon surging to become a $9-billion powerhouse.[32] Its founder, Aniruddha Sharma, won the 2022 "Entrepreneur of the Year" award based on his vision of capturing 1 billion tons of CO_2 by the 2030s. Today, the company boasts some 200 employees and ten partnerships with a range of key industry players across the globe.[33]

While tech incubators are often considered to be an American phenomenon, in fact they can be found all over the world. Climate-KIC Accelerator, sponsored by the EU, is arguably the planet's most sprawling network of incubators, with branches from England and Greece to Egypt and Nigeria—twenty-eight countries in total.[34] To date, 2,100 startups have found a home in a Climate-KIC branch, with 200–300 applicants accepted each year. In 2022, startups supported by the incubator raised €1.1 billion in capital as they moved into the real world.[35]

While the incubator's success stories are too numerous to count, a couple of additional examples provide insight into the rich diversity of firms and their products. While still a Climate KIC client, the Swiss carbon-removal company Climworks was able to raise €27 million and strengthen its position within the industry.[36] The Munich-based company tado° makes smart systems for home thermostats;[37] in 2016, while still based at Climate KIC, the company raised €20 million, which enabled it to expand to US markets.[38]

Alongside governments, many universities have also established incubators. In 2009, for instance, Stanford alumni Tom Steyer and Kat Taylor helped endow the TomKat Center for Sustainable Energy on their old California campus. With a small staff, this extraordinarily fruitful program is designed to tap the creativity of the Stanford community. Interestingly, 30 percent of the winning grantees come from the business school, 60 percent from engineering, and 10 percent from other departments, such as law or sustainability studies.

Brian Bartholomeusz brings authentic climate tech "bona fides" to his job as TomKat Center's executive director, having held senior technology management positions, cofounded three startups, and received fourteen patents.[39] The first stage in the process through which the incubator shepherds fledgling ventures is customer discovery: it is critical to find out a priori if the product is something that customers need and will pay for. "Sometimes people come up with concepts that are 'nice to have' as opposed to 'must have,'" Bartholomeusz observes.[40] Grantees are immediately encouraged to build and test a prototype, for validation: "As quickly as possible, we need to determine if a project has legs, and the best way to determine this is to have the product and business assumptions collide with customers and the market."[41]

In contrast to VC-funded accelerators, the criteria applied during Tomkat's "auditions" are far less "bottom-line" and more public-spirited. Bartholomeusz is proud of this perspective: "We don't waste time on spreadsheets. When we evaluate proposals, the questions are [as follows]: Are you solving a real problem? Do you have a superior solution? Who are your competitors and how might they react? Will customers pay for your solution? And do you have a market entry strategy?"[42] Thus far, during its first ten years of operation TomKat seems to be unusually good at identifying real problems and real solutions: the TomKat Center has awarded just over $6 million in grants to 105 teams, with a total of 255 participants. Of these, eighty companies were launched into a commercial phase, and almost all the new companies remained in the United States.

Economically, TomKat has been a bonanza: all told, companies that it sent into the world have gone on to raise $1.7 billion. These enterprises' collective value is estimated to be far higher, reflected in some two thousand employees at roughly eighty companies, representing $6 billion in generated value—a handsome multiplier by any standard. Aurora Solar, one of the original grantees in 2013, recently achieved "unicorn" status.[43]

The TomKat story is not anomalous. Academic evaluations of the efficacy of university incubators have been going on for decades,[44] and the findings are unfailingly positive.[45] As might be expected, university incubators with for-profit structures have a better track record in spinning off successful companies than do centers that focus primarily on scientific research.[46] As climate innovation has become increasingly common, there is now evaluative research about green incubators . . . and the jury is in: *They work.*[47]

In a comprehensive review, the United Nations Framework Convention on Climate Change (UNFCCC) concluded

> Incubators and accelerators thus play an important and multidimensional role in supporting new climate-resilient and low-emission technologies to be developed, accepted, and used by society. Ultimately, they have the potential to catalyze the development of more-sustainable and inclusive societies.[48]

But the same United Nations report concludes that, as of 2018, of the 2,000 technology incubators and 150 accelerators worldwide, a mere 3 percent were devoted to climate technology. Only a fraction of these are based in developing countries.

While the number of climate tech incubators is growing, it is unrealistic to expect the private sector to be much more involved. Because of the patience and resources required to incubate decarbonizing inventions and infrastructure in the early stages, most VCs remain loath to make the investment, preferring lower-cost programs and faster returns. A few farsighted corporations see the potential for gain and have already begun to sponsor incubators, while philanthropies such as Breakthrough Energy Ventures are already way ahead of the curve. Universities around the world are stepping up. Yet government support will remain paramount in importance, especially to ensure that there is equal opportunity to join the climate tech ecosystem. Anyone would be hard-pressed to find a better way to spend public resources on applied decarbonization research.

Evaluating Climate Research Policy Success

While many activists feel that present government commitments to climate research are woefully inadequate, objective data worldwide confirm that funding directed to decarbonization research is increasing rapidly. The key question is: Will it produce *meaningful* innovation? This turns out to be difficult to answer, and not just because research, by definition, is based on overcoming unknowns. Innovation may be difficult to define—but it is even harder to measure. And yet, the popular policy axiom holding that "if something can't be measured, it can't be managed" surely holds true for research programs. There is no shortage of metrics that can be used to evaluate and gauge the effectiveness of R&D initiatives.

In academia, publication in high-impact, peer-reviewed journals constitutes the touchstone for successful research and the basis for promotion along the tenure track. Most academics know, however, that any connection between prestigious journal publications and invention of a commercial product for reducing carbon footprints is probably coincidental. This explains why patents have emerged as the "gold standard" for measuring innovation. The key role of patents in protecting intellectual property, incentivizing innovation, and spurring economic development is self-evident. Patent applications, in general, are rising as never before. In just two years, between 2019 and 2021, the number of new patents granted in the United States alone went up from 390,499 to the astounding figure of 669,434.

From the perspective of program evaluation, patents have many advantages when trying to measure whether policies (or research programs) are effective in stimulating innovative decarbonizing technologies. Social scientists and economics researchers enjoy easy access to patent-related information contained in sundry databases maintained by intellectual property offices. In theory, patents represent a measure of innovative activity that can be quantified and tracked over time. Presumably, patents are only granted for novel and nonobvious inventions that offer practical and useful applications.[49] Some scholars blithely assume that patents offer an objective marker of innovation *quality*. But it is not at all clear that patents are a good indicator of creativity and transformative technological progress.

Critics argue that patent data are notoriously unreliable and frequently irrelevant as a measure of innovation. That is because most innovations are *never even patented*. At the same time, patent registries are full of inventions that have no practical value, and some are mere oddities. It is also not uncommon for patents to be granted for tiny incremental improvements to already-existing technology. Companies on occasion will tactically apply for patents to prevent competitors from patenting similar ideas.

In *The 9 Pitfalls of Data Science*, Gary Smith and Jay Cordes argue compellingly that, at best, patent data are a very "noisy" measure of innovation:

> Matches are not patented; nor are three-point seat belts, computer mouses, and USB technology. Monoclonal antibodies are not patented, though they comprise six of the ten best-selling modern drugs. Since 1963, the journal Research and Development's R&D 100 Awards have identified and celebrated the top 100 revolutionary technologies of the past year. . . . Only 10 percent of these R&D 100 innovations were patented.[50]

It would seem that patents can be one *indicator* of innovation. But their value should not be exaggerated. Arguments about the impact of research programs that produce numerous patents for climate tech would be far more persuasive if their performance indicators were supplemented by other quantitative measures. These might include the amount of matching funding generated for research and development, the profits produced by the mature products, or the actual amount of carbon reduced.

Effective Climate Research Policy: The Case of Denmark

There are a dizzying number of new technologies that must emerge if the world economy is to transition to net-zero emissions. Scientists need to identify ways to cost-effectively capture carbon, store energy, produce hydrogen, remove carbon from the production of cement, steel, textiles, and plastics, replace meat and dairy products, raise rice without releasing methane, and much more. Given the diversity of disciplines involved, it is difficult to generalize about a monolithic research strategy for decarbonization. Yet, regardless of the specific technology, successful R&D programs share common attributes:

- *Setting clear objectives*;
- *Building on relative strengths—including human capital*;
- *Ensuring sufficient funding*; and
- *Creating a culture of collaboration.*

Countries that have achieved these goals tend to have effective climate technology development programs—and there is no better example than Denmark, which consistently "punches above its weight" in the climate tech space. The country is especially associated with exceptional achievements in researching, developing, and scaling onshore—and more recently—offshore wind turbines.[51] But it seems as if that was just Act I. Notwithstanding the country's modest size, the Danish government has built an ambitious vision for the country's contribution to climate technology.

Denmark's research policy can be seen as an expression of the 2020 Climate Act, when a large majority in the Danish parliament voted to set an ambitious target of 70 percent reduction in greenhouse-gas emissions by 2030 (compared to 1990 levels). Officially, the research program stands on two legs: technological solutions that allow the country to reach its statutory emissions target, and specific research initiatives that will hasten a global green transition.

Available funding is not prodigious: for the foreseeable future, the government is committed to financing climate research at a level no lower than 2.3 billion DKK, or $320 million per year. The Ministry of Higher Education and Science is empowered to oversee the national research strategy which it views as a continuation of Denmark's illustrious green tradition. The ministry waxes eloquent about its mission:

> We can stand on the shoulders of this tradition and develop further. By developing green solutions that can be used in other countries, we can set a footprint that is far bigger than what can be expected, considering the size of Denmark and emissions within our borders. This has already been demonstrated with the development and export of energy technology.[52]

As a small country with fewer than six million people, clearly Denmark can't do everything. Research is expensive and priorities need to be set. In 2020,

Denmark began the process for picking its top climate tech priority areas, weighing each on the basis of the following:

- *Green Potential*, or the extent to which research can provide possible new solutions to the challenges faced by different sectors;
- *Business strengths and potential* refers to technological areas where Denmark already enjoys a competitive advantage and where the world market offers meaningful business opportunities;
- *Research strength*, the third criterion, identifies existing "strongholds" where there is potential to attract additional research funding, especially from EU research programs; and
- *Partnership potential*, which considers the opportunity for cooperation between universities, companies, technology organizations, and other relevant government and nongovernment institutions.

One of the reasons that Denmark's strategy appears so promising is the successful integration of a large and varied cast of stakeholders, including representatives from disparate government ministries, academia, innovation networks, NGOs, public and private funds—and perhaps most important of all, Danish industry.

It can be argued that the relative homogeneity of Danish society contributes to its exceptional team effort.[53] There is a long tradition of major political agreements across party lines (healthcare, education, tax, security and defense, energy), and close cooperation between business and labor unions. Politically, the country's center-left parties were the initial supporters of renewables. Yet it did not take long for right-wing parties to become committed to energy independence. After the Danish flag was burned in several Middle Eastern countries following the publication of editorial cartoons about the prophet Mohammed in a Danish newspaper, they, too, embraced the local climate tech ecosystem.[54]

Some Danish experts track the county's commitment to renewable energy all the way back to 1973, when Denmark suffered huge inflation and unemployment because of soaring oil prices. The government responded by taxing fuels and subsidizing wind development—and did not change course even when oil prices began to fall again.[55] Consensus and community spirit have always been highly valued across Danish society, and the country's global leadership in wind energy soon became a source of national pride, irrespective of political ideology.

In recent years, the government established no fewer than fourteen formal "climate partnerships" with the country's Green Business Forum (composed of cabinet ministers, business and labor leaders, and independent experts) to work together on additional technological challenges. The partnerships include businesses involved in everything from aviation, shipping, defense, and agriculture to construction, manufacturing, and finance.[56] The government sets long-term goals, but it relies on the private sector to supply the investments required to get there. As part of this dialogue, some 400 recommendations have emerged, informing the country's research priorities.

When a country is involved in a strategic planning process, it makes sense to be fully cognizant of what other competing countries are doing. Fittingly, in 2019 the Ministry of Higher Education and Science convened a panel of international experts. The panel concluded that Denmark should go beyond the prevailing approach of simply financing stand-alone projects and adopt larger, coherent, long-term efforts. Accordingly, at least 750 million DKK each year is dedicated to

- Capture and storage or use of CO_2;
- Green fuels for transport and industry;
- Climate- and environment-friendly agriculture and food production; and
- Recycling and reduction of plastic waste.[57]

With the overarching goal of accelerating "a green transition," the policy sought to link the country's research agenda to the actual interests of commercial Danish companies. In fact, the Danish government had a head start in climate R&D because it had already streamlined public research financing in general. In 2014, a motley assortment of small research institutions were merged into a single Innovation Fund Denmark (*Innovationsfonden*) that was then granted independent agency status.[58] The mission of the fund is to promote Danish innovation and economic growth through support for research and development. Climate tech is hardly the only business sector it supports, but the government decided to make it among the most preferred.

While the fund's mission is to create growth and employment, it has fundamentally different values from those of private investors:

Compared to other investors, Innovation Fund Denmark's results do not necessarily need to be seen in share prices or end-of-year financial results. Innovation Fund Denmark's results must also be evaluated on social welfare improvements, increased societal wealth, jobs, reduction of CO_2 emissions, and a cleaner environment.

It also prioritizes research projects that have "high risk—and also high potential . . . with as little bureaucracy as possible."[59]

One example of Denmark's ambitious research program is the prioritization of carbon capture use and storage (CCUS). All roadmaps to decarbonization recognize that even the most effective transition to a low-carbon economy will not eliminate emissions entirely. Under the most optimistic scenario, 10 billion tons of heat-trapping gasses will still be released on a global scale annually. Without technologies to remove them, net zero will remain elusive.[60]

The problem is that until carbon emissions are monetized there will not be a market for the technologies that need to be deployed. This is especially true for the foreseeable future, because these technologies are so expensive. If carbon capture is going to be used widely, it must get cheaper, and this will not happen without applied research. In what can be seen as a very long-term investment, Innovation Fund Denmark decided to double down on CCUS technologies.

Karina Søgaard, a bio and chemical engineer, is the partnership director for Denmark's CCUS program. Søgaard cobbled together a heterogeneous leadership team of scientists and industry experts to tackle the problem. At present, she oversees a five-year budget of €100 million, which has already given rise to forty-nine demonstration projects designed to last between one and four and a half years. Most of these fall into the category of public–private partnerships, collaborations that combine the strengths and resources of both corporations and the government. Both sides invariably feel as if they have leveraged the other.[61] That's a good thing.

The Danish CCUS's collaborative programs could be poster-children for government-sponsored research programs: two-thirds of Søgaard's budget comes from government funding and one-third from the private sector. The CCUS research program is overseen by a ten-person board of directors: five from the research sector and five from industry. On the whole, politicians and even government officials do not intervene in the decisions of the partnership. Instead, the Innovation Fund expects regular reports but allows Søgaard and her team to work unhindered. Søgaard candidly recognizes that the research strategy's initial version was "imperfect," but says it still offered realistic goals and was good enough to move forward and improve as experience accrued.

Denmark's climate tech research partnerships are specifically designed to support industry in developing technologies that are located between stages 3 and 7 in the technology readiness level (TRL) scale. (Typically, knowledge institutions like universities contribute at a technology's early stage—a few years before it is "market mature"—or calibrated at TRL level 5.)[62] This means that the CCUS program aims to strengthen companies who are working to provide experimental proof of concept and to support validation in the laboratory, as well as fund projects that expedite successful demonstration in the operational environment.[63]

The CCUS research program did not hesitate to start big: it immediately sought out the country's single largest CO_2 emitter, Aalborg Portland, the only gray- and white-cement producer in Denmark.[64] At present, researchers sponsored by partnership funding are testing how to capture CO_2 from the flue gas.

The CCCUS program is just one facet of Denmark's ambitious plan to be a significant player in the clean tech market, a goal that ultimately depends on human ingenuity. In that respect, Denmark is well-poised to innovate: it is a highly educated country, with over 30 percent of the Danish population completing some form of higher education.[65] The government is committed to supporting "the coherence between our study programs and the green transition."[66] But since Denmark is also a country with extremely high taxes (individual income tax is often well over 50 percent),[67] brain drain is an increasing concern. Some of Denmark's best-trained workers opt for an expatriate status in other countries that can offer them substantially higher earnings. But most opt to stay, enjoying their role in crafting and implementing public policies that maximize returns on an impressive climate tech research strategy.

ReSource's Path Across the Valley of Death

Plastic, once heralded as a miracle substance, is today regarded as one of the world's most intractable environmental problems. Its disposal is a major issue, but so is its manufacture: since plastic is made from oil and gas, its production is a major contributor of greenhouse gasses. If plastics could be designed using CO_2 from the atmosphere as raw material or produced indirectly from plant biomass, then the industry could join the circular model of the world's future net-zero economy. Unfortunately, creating new carbon–carbon bonds that build useful organic molecules is a process which requires energy-intensive reactions. As a result, until now, any environmental advantages from using CO_2 as a feedstock tend to be undermined by the chemistry. In other words, the CO_2-utilization processes emit more CO_2 from energy consumption than are utilized in the process itself.[68]

Since he established his chemistry laboratory at Stanford University, Matt Kanan and his graduate students have explored possible solutions to this problem, synthesizing different chemicals using CO_2 as an alternative to conventional petrochemical molecules. Funding for the lab's work came from the usual suspects: grant-making agencies in the federal government and organizations that fund chemistry research.[69]

In 2009, Aanindeeta Banerjee joined Kanan's laboratory as a doctoral student and began to devote her attention to creating new ways to convert CO_2 into more-usable building blocks for chemical reactions. It would take a full seven years for her findings to be published in *Nature*, the gold standard for important scientific discoveries.[70] Working with Kanan and her colleagues, Banerjee discovered a reaction that inserts a CO_2 unit within an existing carbon-hydrogen bond in a different organic molecule. Even without using expensive catalysts or organic solvents, the process can be used to form carbon–carbon bonds and construct more complex molecules such as a molecule called FDCA.

FDCA is the monomer from which PEF, a potential replacement for PET plastics, is made. In other words, the novel chemical reaction has the potential to replace conventional plastics and simultaneously sequester atmospheric carbon. From a climate mitigation perspective, it's a grand slam. Rather than continue her career as an academic, Banerjee decided that she would devote her energies to transforming this discovery into a marketable product. Along with Kanan, she founded a new company, ReSource Chemical Corporation, and, as CEO, she became its first employee.[71] Banerjee and her company's journey offer an excellent example of how American research policy in the climate space is working and how it could yet be improved.

Still just a very good idea, one backed by a few scientific publications, ReSource Chemical received a $50,000 grant from Stanford University's TomKat Center for Sustainable Energy to get the ball rolling. The money allowed ReSource to hire Ken Keckler, a chemical engineer with considerable experience working at British Petroleum. It took a year for Keckler to develop a detailed model that could translate ReSources's chemical reactions into an industrial manufacturing process. This work made Banerjee much more confident about the company's commercial viability and the many environmental advantages it could offer.[72]

Banerjee was a natural candidate to receive an "Activate Fellowship" for professional development through the University of California Berkeley's accelerator, Activate Berkeley. The center adopts promising new ventures in so-called hard tech—harnessing both scientific and engineering expertise to tackle difficult technological problems. The fellowship provided a stipend, along with $100,000 in grant money for research equipment and material costs. It also introduced Banerjee to a community of mentors and previous cohorts of aspiring entrepreneurs. The training allowed her to take the next step in her makeover from lab scientist to corporate CEO. It also provided the connections that would allow the company to set up shop at the Lawrence Berkeley National Lab from 2020 to 2023. There, she was joined by another brilliant chemist from the Kanan Lab, Dr. Amy Frankhouser, who stepped in as vice president for chemistry.

ReSource has the potential to be a classic Silicon Valley climate tech Cinderella story. But climate tech is not high tech like cyber-security software or game apps. It takes a long time. By Kanan's estimate, it may be twenty years before an industrial production facility can begin to show the world the economic and environmental benefits of cleaner plastic. The company continued to scramble for support on its long journey, receiving a $250,000 National Science Foundation grant in 2020. The US Department of Energy also understood the company's potential and selected ReSource as a finalist to receive a significant $2-million grant. Unfortunately, it was contingent on a matching contribution of $500,000.

This was a potentially game-changing stage for the company, one that would give ReSource the financial capacity to buy equipment for scaling up its process. But government agencies move slowly. It would take almost a year to find out whether the DOE award would actually be made, and then another five months to negotiate the details. The funds were not provided up front but only distributed on a reimbursement basis. In practice, this created serious liquidity challenges as the company had to wait for several months before it finally received money it had already been awarded. In 2023, ReSource would apply for another million-dollar grant from the National Science Foundation, which this time could be provided in four up-front payments. VC support has provided the company with some additional oxygen to move forward. With faith in the significance and uniqueness of their process, and fortified with the support of their Stanford doctoral advisor, Banerjee and Frankhouser soldier on.

ReSource enjoys many unique advantages. It benefits from the scientific acumen and oversight of an academic superpower. It was born into the world's most dynamic and prosperous entrepreneurial ecosystem. It has a particularly gifted and well-motivated team made up not just of world-class chemists, but also of individuals with the audacity to go beyond the limits of laboratory science. Most of all, ReSource has a breakthrough technology with the potential to transform a carbon-intensive industry that is a scourge for the world's oceans and a global solid-waste menace as well.

Even with all these advantages, the company's future—and the crucial transformation it might bring to the planet, are in no way assured. But it is clear that without a national policy prioritizing climate tech research, ReSource will have had little chance of surviving. Because public support was available early in the process, the company may be able to attract the necessary venture capital to go the distance. But its circuitous path and long time horizon suggest that a more aggressive policy, with far greater economic muscle, will be required if the present fossil-fuel-driven economy is to evolve into a sustainable one.

The Imperative of Effective R&D Policies

In 2021, an important article written by researchers at Cambridge University appeared in the distinguished journal *Nature Climate Change*. The authors, Cristina Peñasco, Laura Anadón, and Elena Verdolini, reviewed 211 scientific publications that evaluated the effectiveness of policies designed to mitigate climate change. Many of the policies are the same ones described in this book: renewable feed-in tariffs, portfolio standards, emissions-allowance trading schemes, public procurement, carbon taxes, tax exemptions, and of course, research and development funding.

The Cambridge scholars found strong agreement that *all* policy measures had a positive impact on environmental, and to a lesser extent, technological progress.[73] Some were better than others. And the authors also found that public expenditures on R&D improved competitiveness, attracting venture capital funding for the cleantech sector. No policy appears to stimulate innovation as effectively as direct funding for research and development:

> Recent literature indicates that the design of R&D funding schemes can help foster firm-level competitiveness outcomes beyond the positive impact on innovation outcomes. Government R&D funding programmes targeting small companies or those in early stages of development help attract other funding sources and advance small-firm competitiveness. . . .[74]

The message is clear. If society wants to accelerate climate tech, it must be willing to invest in research and development. There are no guarantees. And there are no shortcuts. Yet global experience suggests that a significant investment can deliver the technologies so critical to living sustainably on planet Earth—if it is backed by smart research policies.

CHAPTER 4

Monetizing Carbon

In February 2009, Michael Greenstone took over as chief economist at the Council of Economic Advisors for the newly inaugurated president, Barack Obama. Just ten years out of graduate school, Greenstone was already a tenured professor of economics at MIT and was recognized as a leading scholar on environmental matters. He arrived in Washington with visions of helping to craft a national market for carbon as an efficient way to ratchet down the country's greenhouse-gas (GHG) emissions.

Unfortunately, the shockwaves from the bursting of the American housing bubble in 2007 were still raging, following a rash of mortgage delinquencies and foreclosures around the country. The US economy was in free fall, with 700,000 people a month losing their jobs and a new, untested president expected to pick up the pieces. Several earlier legislative proposals— even some bipartisan bills— seeking to create emissions trading systems or impose carbon taxes had never gotten off the ground in Congress.[1] It was clear that a new energy tax would have *zero* political traction. And yet, Greenstone had come to Washington because President Obama was deeply concerned about the climate crisis and wanted to do something to address it.

Greenstone decided to devote his considerable energies to "Option B": requiring the federal government to consider the full cost of greenhouse-gas emissions when making regulatory decisions. The problem, as Greenstone saw it, was that the deck was stacked against climate mitigation: "What you had on the one hand was the price of any restriction that imposed costs on industry and consumers—as opposed to the benefits, that were only referred to vaguely as 'tons of CO_2.' That was clearly an unfair fight. The money was always going to win."[2] As an environmental economist, Greenstone knew that those amorphous "tons of CO_2 emissions" represented real costs. They just had to be converted to dollars.

Greenstone recalls the prevailing dynamics in the federal government at the time:

> I had this misconception that each branch of the federal government was trying to maximize social welfare. But then I looked around and immediately saw that wasn't true: every agency was trying to push its own agenda. It's not that they weren't aware of a broader social context, it was just secondary for them. The Department of Transportation basically wanted more people driving, more miles, as if there were no associated environmental costs. But at the same time, the EPA treated carbon as if it had infinite costs. I thought that at a minimum, every agency should price emissions in the same way.[3]

Greenstone met Cass Sunstein, then serving as administrator of the Office of Information and Regulatory Affairs in the Office of Management and Budget, over lunch. He convinced Sunstein that without setting a uniform price on carbon emissions, consistent across all government agencies, federal climate policy would be inconsistent and little progress would be made. The two established an interagency process, using the leverage of the White House on government agencies, to adopt a "social cost of carbon." The price was to reflect the marginal cost of every additional ton of CO_2 released. (An alternative, positive framing would hold that it measures the marginal *benefit* of reducing carbon emissions by a ton.)

With climate legislation a nonstarter on Capitol hill, the Interagency Working Group on Social Cost of Greenhouse Gases proved to be the only game in town for people in Washington, DC, interested in moving the needle in national climate policy. But creating a consensus between government bureaucracies with such contradictory perspectives turned out to be a particularly vexing task. Setting the price was not just a question of understanding climate models and characterizing anticipated damage—projections that are not without considerable uncertainty. There are also valuation problems. It is not self-evident what price should be put on health, such as increased cases of heatstroke. Or what about foregone recreational activities? Innumerable social dilemmas are also germane: Should the United States only count damages taking place internally, or consider global impacts? How should the discount rate be set? (Or what value should be set today for cash flows in the future?) In other words, how much do people care about their grandchildren's future?

In retrospect, Greenstone calls forging a consensus between the different departments the hardest thing he has ever done in his life. But he did it. In February 2010, the Obama administration published Executive Order 12866, containing the first formal Social Cost of Carbon estimate, using global criteria. It set the price at $36 per ton of CO_2 equivalent (CO_2e) in 2015; $42 by 2020; and $46 per ton by 2026.[4] In 2009, the US EPA started using carbon pricing in its cost–benefit analysis for vehicle emission standards. The approach began to spread: the states of Colorado, Minnesota, and Washington all started requiring electric utilities to use the federal price of carbon in their resource planning.

Many environmentalists believed that these prices were set too low. For instance, if society truly cares about the future, it should eliminate a discount rate altogether. When this happens, the full force of climate-change-driven ecological damages would be counted centuries ahead. Such damages should include massive sea-level rise, loss of food-growing capacity, etc.[5] Using that assumption, one recent study sets the price from $10,000 to $750,000 per ton of carbon.[6] Conservative economists see such figures as fanatical.

Soon after taking power, the Trump administration decided to eviscerate Obama's executive order, limiting the relevant climate impacts to those within the United States. The carbon price tag immediately plummeted tenfold, to between $3 and $5 per ton.[7] But the pendulum swung back: when President Biden was elected, he wasted no time in resetting an interim carbon value at $51, a number which was raised in 2020 to $190 per ton, going up to $230 in 2030.[8]

Governments of other countries have begun to use different figures when applying a social cost of carbon. Formally, the calculations tend to follow four steps: first, future emissions are predicted based on expected economic and population growth; second, the environmental effect of the emissions scenario is modeled; third, the economic impacts that climate change has on agriculture, health, property values, biodiversity, energy use, and other aspects of the economy are calculated; and finally, future damages are converted into present-day values and aggregated to determine the total price tag.

The problem is that, notwithstanding the pretense of objectivity, estimates are frequently informed by political considerations rather than strict economic calculations. For instance, the federal government in Canada decided in 2019 to apply a social cost of $50 per ton of CO_2 equivalents. At the same time, it announced that this number was already too low and did not reflect the actual cost for society. Hence Canada would increase carbon prices by $15 per year starting in 2023—until reaching $170 per tonne in 2030.[9]

Greenstone left his government position in 2010 after President Obama's executive order was released and was soon recruited by the University of Chicago, where he set up its Climate Impact Lab and continues to improve the empirical estimates of climate change's impact. "We've been working on it for a decade and some surprising things have emerged." He explains,

> The damage function is much more nonlinear than we had previously thought. Thresholds get crossed in a wide variety of human wellbeing indicators: mortality, unpleasantness at work, crop yields. When you don't have data, you treat wide regions of the world as homogeneous. In the first estimates, the United States was treated as a single region. That assumes that the effects in Minnesota would be the same as in Houston or Miami. Once you increase the quality of the model, what emerges is incredible inequalities in climate impacts. Some places will be fine. Some places will be harmed. And some places will be really, *really* harmed.[10]

For example, Greenstone notes the differences in human mortality in various parts of the world. It is often argued that any increase in deaths due to heat waves will be balanced by the lives saved by more-moderate winter temperatures. But this falsely assumes that poor people in the Global South are as capable of protecting themselves from the extreme heat as people in wealthier Northern lands are of evading the winter cold. With the power of high-resolution models, Greenstone's research is injecting a critical element of climate justice into carbon monetization.

It is little wonder that many analysts continue to challenge the precision of the metric and propose alternative approaches.[11] But there is a consensus that both carbon taxes and emission trading systems suffer from excessively low prices on carbon. How much, then, *should* society charge firms—and individuals—for greenhouse-gas emissions caused by their activities? On a theoretical level, at least, carbon markets should reflect the actual harm caused by emissions. Unfortunately, thus far, they have not come close.

Carbon Taxes: The Great Disappointment

Many environmentalists have long argued that the most cost-effective way to reduce pollution is to simply regulate it. Set a limit on carbon, and companies will be forced to stay below that limit. But command-and-control rules ultimately do not motivate companies to go above and beyond, or in this case, below and beneath: polluters tend to stay right below the legal limit but do not reduce emissions further.

By contrast, if you set a price on carbon, the less emissions the company produces, the less it must pay. The theory is straightforward and compelling: money is a motivator. Economist David Popp had long posited that, if the incentives are right, self-interest will induce firms to innovate climate-friendly technologies.[12] There is also the moral logic of Pigouvian "sin taxes": society should make products that are good for society, less costly; and those that are harmful, more expensive.

That's why, for many experts, the go-to policy for reducing emissions has always been carbon taxes—or emissions trading systems (ETS). Emissions trading has the theoretical advantage of guaranteeing that a "cap," or the maximum acceptable level of emissions, is not exceeded. For many years, such policies were the hypothetical musings of environmental economists. No longer. The remainder of this chapter considers how these two competing approaches have fared in the rough-and-tumble of the real world.

Carbon taxes have actually been around for quite a while, so there is no shortage of data for evaluating how effectively they perform. It turns out that, empirically, they don't seem to work very well in reducing emissions. And even more disappointing, they hardly seem to catalyze innovation at all.

British Columbia is frequently singled out as the poster child for carbon taxes. Enacted in 2008, the BC policy was done by the book, conforming to all the experts' sage recommendations. Due to unusual resolve on the part of Gordon Campbell, British Columbia premier, the tax was adopted without granting excessive exemptions to politically influential industries. The tax started low, at only C$10 per ton of CO_2, with a mandate to gradually increase it by five dollars each year, as the public became accustomed to the new levy. This avoided the angry political pushback that would later greet—and soon defeat—a moderately pricey Australian carbon tax in 2011.[13] By April 2023, after steady, iterative hikes, the BC tax hit $65. This translates at the pump to an increase of 14¢ per litre of gas, 17¢ per litre of diesel, and 12¢ per cubic meter of natural gas.

A key reason the government was able to overcome political opposition to the tax was that it was "revenue neutral." Led by British Columbia's ruling pro-business Liberal Party, the government promised that any money collected under the tax would be returned to citizens, particularly small businesses, rural areas, and poor communities. As a show of good faith, Campbell's finance minister, Carole Taylor, even announced that she would reduce her own salary by 15 percent if she did not deliver the revenue-neutrality as promised.[14] The government was as good as its word. Over half of the tax revenues were balanced by parallel cuts in individual and corporate income taxes.

Initial reports about the effect of the tax were effusive. One Canadian think tank waxed ecstatic:

> BC's fuel use is down a whopping 16.1 percent. Its economic growth has kept pace with the rest of Canada. And its personal and corporate income tax rates are now among the lowest in Canada. In short, the numbers indicate that BC's carbon tax shift has been a remarkable success, environmentally and economically.[15]

The law really did appear to relieve the tax burden for underprivileged and low-income taxpayers; advocates could justifiably frame British Columbia's carbon tax as a progressive policy.

After several years of implementation, Yale University economies professor Stewart Elgie offered a highly positive evaluation of the carbon tax:

> What British Columbia has done is lowered taxes on investment income and employment, and increased taxes on pollution, which is exactly what we should be doing with our tax system. Pretty much every respected economic body in the world in the last few years has said we're moving toward a global economy that will reward low-carbon, innovative, resource-efficient production. And if we don't prepare ourselves for that, other countries are going to eat our lunch.[16]

Subsequent reports, however, have been less enthusiastic in their appraisals. While a few indicators are encouraging, recent data about actual carbon emissions are far less sanguine. British Columbia's 12-ton average per capita CO_2e emission rate is certainly 32 percent lower than that of other provinces in Canada. But it is still 300 percent higher than average individual emissions in Sweden.[17] Sadly, the impressive initial reductions, which were consistent with economists' optimistic projections, did not continue. Recent government reporting from British Columbia admits that per capita GHG emissions have risen in recent years, wiping out any modest improvements attributed to the carbon tax that took place in previous decades. By 2018, total greenhouse gas emissions in British Columbia were up to 67.9 million tons of carbon dioxide equivalent (Mt CO_2e). This represented a jump of 3.3 percent in a single year and a 7.1 percent increase in gross emissions since 2007, British Columbia's baseline year for assessing reductions in greenhouse-gas emissions.[18]

Felix Pretis, from the University of Victoria, published a comprehensive assessment of the British Columbia carbon tax and was candid in expressing his disappointment in its performance. Pretis's conclusions were based on a "difference-in-difference" study. This methodology is used to characterize causal relationships in government intervention programs where randomized controlled trials are impossible. In this case, Pretis compared the emissions measured during the course of the carbon tax program in British Columbia with other "control groups": ten other parallel Canadian provinces that chose not to impose a carbon tax. Besides a modest drop in emissions from mobile sources in British Columbia, he saw no real difference in the aggregate CO_2 emissions. In fact, other provinces that *did not even* have a carbon tax, such as New Brunswick and Nova Scotia, actually reduced emissions more than did British Columbia. The drop in greenhouse-gas emissions in New Brunswick is associated with the closing of fossil-fuel-powered generators, and in Nova Scotia, the shutting down of pulp and paper mills.

There is even a possibility that one of the main achievements reported about the British Columbia carbon tax was misleading. Rather than catalyzing a wave of electric vehicle purchases, a 2018 University of Ottawa doctoral dissertation provides strong evidence that the British Columbia carbon tax may have actually engendered unintended consequences: substitution of gasoline-powered cars with diesel vehicles. This was soon reflected in increased pollution of British Columbia's air.[19]

Sadly, it's the same story all over the world. Professor Johan Lilliestam and his formidable energy policy team at the Helmholtz Centre in Potsdam, Germany, reviewed the effect of myriad climate taxes adopted around the world. By the year 2020, fifty-seven disparate carbon pricing initiatives were in force. The earliest taxes were enacted in quick succession in Scandinavia, with Finland, Norway, and Denmark levying carbon taxes in 1990, 1991, and 1992, respectively.

The experts evaluated all peer-reviewed research about carbon taxes internationally and concluded that they produced only small emission reductions. (The modest drops were due to people switching the kind of fuel they used and using slightly less gasoline.) The research team decided that there were basically "no effects on low-carbon investment and innovation and no effects on zero-carbon investment."[20] The Cambridge review of climate policies described in chapter 3 came to a similar conclusion: it analyzed twenty-seven studies about carbon taxes and found no clear, measurable effect on innovation. Worse, 63 percent of the publications reported negative impacts on social equity. For instance, several studies concluded that rural areas were hit harder by energy taxes than urban areas.[21]

What happened?

When experts consider the impact of carbon taxes, two flaws are usually blamed for the disappointing results: taxes are set too low to influence behavior, and too many key industries are given exemptions from paying carbon fees. Regarding the former, politicians have been decidedly meek in setting the price of carbon. It is not at all clear how high a tax needs to go to effect change. For example, Sweden has the steepest carbon tax in the world—more than twice the rate of the much-hailed British Columbia levy. Although Sweden started in 1990 at a modest rate of $30 per ton of CO_2, it was gradually raised. Today the tax sits at $126 (€118) / CO_2 ton.[22] Even so, two evaluation studies, conducted in 1998 and 2018, reached the same conclusion: the tax had no significant effect on CO_2 emissions.[23, 24] Sweden's carbon footprint has surely diminished. But econometric data suggest that the decline was due to the growth of nuclear and hydro power and other energy policies, along with net electricity imports from other countries. It was probably *not* a result of any change in practices due to the carbon tax.

As for overly generous exemptions, Sweden again serves as a good example. Julius Andersson from the London School of Economics published an evaluation of Sweden's carbon tax in 2019, and his findings were only slightly more encouraging than those of earlier reviews.[25] It seems that the Swedish carbon tax has only affected transportation, which makes sense given that 90 percent of the tax revenues come from the surcharge on gasoline and diesel. Industry and agriculture pay far lower rates due to the generous exemptions they enjoy. These exemptions were granted due to concerns about competitiveness and carbon leakage. Fuels for electricity production, for example, are fully exempted. And only about a third of heating in Sweden relies on fossil fuels anyway.

Even Andersson's "favorable" review found that annual carbon emissions only dropped significantly, by 11 percent, in just a single year. Moreover, this occurred during a period when European vehicles were becoming more fuel-efficient anyway.[26] By contrast, Norway did not tax carbon directly but instead provided incentives for electric vehicles. It achieved better results through a suite of measures: exempting zero-emission vehicles from registration taxes, VAT, and motor fuel taxes; a 50 percent reduction in road tolls; and discounted ferry prices and parking fees for EVs. These perks appear to have been highly effective: by

September 2023, 93 percent of the cars purchased in Norway were plug-in electric,[27] more than twice the rate of Sweden.[28]

There are other commonsense explanations for the poor performance of carbon taxes. For instance, with fossil fuel prices so volatile, the small difference in price made by a carbon tax would hardly register in an individual consumer's cost–benefit equation. After the Russian invasion of Ukraine, gasoline prices worldwide went bonkers. Canada was no exception.[29] On March 2, 2022, the average cost for Canadian gasoline was C$1.55/liter. By June, the price had spiked at C$2.09. And then, six months later, at the end of 2022, prices at the pump had tumbled down to C$1.37—a drop of roughly 35 percent.[30] With such fluctuations, British Columbia's 17¢/liter increase in gasoline price due to its carbon tax becomes mere background noise.

Pigouvian taxes, it seems, have a better chance of influencing consumer behavior or catalyzing research and development when they are placed on goods with stable prices, like tobacco or alcohol. The 85 percent tax on cigarettes in Bosnia or in Israel are costly enough to get a consumer's attention. Indeed, recent econometric data suggest that every 10 percent increase in Israel's cigarette tax caused a parallel 9 percent drop in smoking rates.[31] Thus far, carbon taxes don't even come close.

Boston University finance professor Nalin Kulatilaka explains that to be effective, carbon taxes must not only influence consumer behavior but change companies' calculus. People cannot buy low-carbon products (such as fuel-efficient cars) if companies have not yet produced them. For those corporations, switching to a more socially responsible technology may be more expensive than simply paying the tax.[32]

Carbon tax maven Johan Lilliestam suggests another, simpler, reason for climate taxes' failure: the absence of alternatives:

> We have a carbon tax on heating in Germany. It is low now, but it will
> be increased soon to 60 Euros. I rent my apartment in Berlin. There is
> no way I can move to a building with a more efficient heating system.
> There is simply no apartment to move to. Already fuel is taxed quite
> heavily across Europe. Adding another 30 cents per liter won't change
> people's actual reality. Unless there are better public transportation
> options, people will continue to drive.[33]

Common sense (and empirical research) suggest that when a reasonable alternative is available, a Pigouvian tax can be highly effective if it is set at a level that can influence consumer demand. A 2016 Oxford University study concluded that a 40 percent tax on beef and a 20 percent tax on milk would be sufficient for people to "internalize" the actual environmental costs associated with their production.[34] That's a mighty sum, but from a mitigation perspective, perfectly logical. The United Nations Food and Agricultural Organization consistently estimates total emissions from livestock at about 14.5 percent of total anthropogenic greenhouse-gas emissions.[35] As protein substitutes, plant-based milks, and cultured meat offer increasingly compelling alternatives to beef and chicken, a carbon tax on animal products today might be more effective in changing consumer preferences than one on transportation, where alternatives may be unavailable. Given the new items on the grocery shelf, it is surely worth a try.

Emissions Trading Systems

Economists have long argued over whether "cap-and-trade" emissions programs are better at reducing greenhouse-gas emissions than carbon taxes.[36] These programs set a limit or cap on overall emissions (e.g., tons of CO_2) and then assign allowances to polluters, monitoring them via a registry. Once firms purchase an allowance, they are authorized to sell it —or buy additional credits if they need more. At least in theory, the system provides a powerful economic incentive: once it becomes cheaper to reduce emissions than to buy credits, the companies will cut back. No less important, once R&D departments come up with emission-control technologies that cost less than the price of carbon, the companies will happily sell their credits for a profit.

Proponents of trading programs argue that they ensure emissions will not exceed a certain level and that once an initial cap is set, it is possible—yea, even essential—to gradually ratchet it down. Pollution control, typically a headache and an annoying expense for companies, suddenly becomes a potential corporate profit center within firms, leading to additional resources for R&D departments to come up with innovative pollution-control solutions. Unfortunately, economic theory often comes up short when it confronts political, psychological, bureaucratic, and engineering realities, which can have a logic of their own.

Most cap-and-trade programs are based on the pioneering, twentieth-century, American acid rain program. Fish in pristine lakes in the northeastern United States were dying because rainfall containing sulfuric emissions from power plants in the American Midwest were driving the pH in the water down to uninhabitable levels. Under the policy adopted by the government, power plants were assigned sulfur dioxide (SO_2) emissions allowances, which they were then able to trade in a government-supervised market. Electric utilities responded to the new incentives immediately, and the trading program functioned exceptionally well. Between 1990 and 2004, a 36 percent *decrease* in SO$_2$ emissions from electric power plants was measured, even though electricity production *expanded* by 25 percent during the same period.[37] Environmentally, it was a triumph: acid rain no longer threatened streams, lakes, and marshes in the northeastern United States. Aquatic habitats rallied and rebounded after sulfate levels dropped by 68 percent.[38] Moreover, there seemed to be no economic downside.

In the early days of the United Nations Framework Convention on Climate Change (UNFCCC), the acid rain program was seen as proof of concept for adopting a global greenhouse-gas emissions trading system. But while there was considerable hoopla about the program outlined in the Kyoto Protocol, ultimately the *global* trading system never really got off the ground.[39] Nonetheless, the protocol *did* inspire the European Union to launch the first greenhouse-gas Emissions Trading System in 2005, creating a system that continues to operate to this day.[40]

The European Union (EU) trading system is relatively broad, including upwards of 11,000 industrial plants and power stations in thirty-one countries. This represents a full 50 percent of greenhouse-gas emissions released on the continent. But it is hardly the only cap-and-trade model. New Zealand, South Korea, Tokyo, Switzerland, California, Québec, China, and the northeastern US states soon followed with their own cap-and-trade programs.[41]

While emissions trading programs cannot be called failures, they also have not delivered the slam dunk hoped for by climate policy enthusiasts. A research team at Tufts University evaluated the performance of eight different greenhouse-gas trading systems already in operation based on five criteria: *environmental effectiveness, economic efficiency, market management, revenue management,* and *stakeholder engagement.* Overall, the study did not find meaningful improvements. The exception was rare praise for a joint program between California and Québec, which covered all the major sectors of polluters (including transportation) while tightening the emissions cap by 3 percent every year.[42]

The most common flaw plaguing cap-and-trade systems, as in carbon taxes, is their narrow scope. Key polluters manage to dodge the system through liberally bestowed performance discounts. The states along the eastern seaboard of the United States, for instance, adopted a minimalist program: their auctions and allowance requirements only apply to fossil-fuel-fired power plants, sized 25 megawatts (15 megawatts in New York) or greater.[43] This amounts to as little as 20 percent of their overall greenhouse-gas emissions. Similarly, the New Zealand emissions trading system only covers about 43 percent of the country's carbon emissions, exempting agriculture and release of greenhouse gases from land use.

Transportation remains unregulated. By way of contrast, when the California and Québec coordinated emissions program moves into its second compliance period, some 85 percent of the carbon emissions in their economies will be regulated under a cap.

New Zealand's cap-and-trade program also suffers from the second-most common flaw afflicting carbon-emissions trading programs: low prices that do little to motivate improved performance, much less catalyze technological innovation. From the outset of the program, carbon prices for New Zealand's local industries have always been low, rarely exceeding US$50 per ton of CO_2.[44] It should be no surprise that empirical studies have shown that the programs has little to no effect on actual emissions.[45]

These examples illustrate a common story: in order to overcome opposition when getting an emissions trading system off the ground, policymakers grant too many exemptions, grandfather in too many polluters, and keep the price of carbon too low. The political logic is clear. In the short term, by reducing the number of "losers" created by emissions trading, pushback is avoided. It also is considered a sound strategy to prevent "leakage" of emissions—or the relocation of pollution sources to more-lenient regulatory jurisdictions as firms seek a path of least resistance. (Leakages also occur when a country simply allows the import of high-carbon-embedded products from uncapped countries.)

The outcomes of such pragmatism have been underwhelming. Stated more bluntly: *No pain, no gain!* Grandfathering in existing emissions as the basis for trading systems leads to a cap that is simply too high to create an allowance-scarcity sufficient to push carbon prices up to a level that spurs meaningful mitigation, much less innovation. In short, if emissions-reductions programs are going to change corporate and ultimately individual decisions, caps need to be steadily ratcheted down over time and more industries included. Regulatory agencies should make it clear from the inception of trading programs that any free allocation of emission allowances will not be permanent. To create incentives for companies to produce more-efficient technologies, emissions trading programs will need to auction off emission credits. This sends a message to participating firms to begin investing in climate innovation.[46] These lessons are important for all trading programs but can be clearly demonstrated by the EU's experience. Because it is the oldest and still the most extensive carbon market in the world, its evolution is instructive.

The EU's ETS: Trial and Error

Europe's emission trading system (ETS) celebrates twenty years of operation in 2025. All twenty-seven EU countries continue to take part in it, along with voluntary participation by Norway, Iceland, Lichtenstein, and the United Kingdom. After establishing a collective—and national—cap, the countries prepared national allocation plans that specified the actual allowances granted to the polluters regulated under the system. These plans let many types of industries off the hook, under the assumption that these polluters would be regulated domestically through other policies.

When the program began, national plans were approved by the EU Commission, and ultimately well over 11,000 major sources of carbon emissions were governed by these rules. A small fraction of the total allowances is retired every year, as the countries seek to incrementally move toward their emissions targets. For instance, the 2020 cap was set 21 percent below the emission levels of 2005, and the overall number of emission allowances is now set to decline at an annual rate of 2.2 percent.

It is common to divide the experience of the EU system into four different trading periods. The first two trading periods paralleled the first commitment period of the UNFCCC Kyoto Protocol and continued until 2012. Unaware that the UN global carbon market would ultimately fail, the EU system was designed to meet the provisos of the Kyoto trading scheme. The third trading period ran until the end of 2020; the fourth is supposed to continue until December 2030.

Initially, allowances were handed out for free, along with the stipulation that factories monitor and report their CO_2 emissions to the European Commission and the Emissions Trading Registry. Firms with leftover credits were allowed to sell them—while those who needed more allowances could purchase more, privately. This was done either by moving allowances within a company, using a broker who matches buyers and sellers, or by trading on the spot market of one of Europe's climate exchanges.

Is it working? Looking back, it is hard to say whether Europe's ETS program has been a force in reducing greenhouse gases. GHGs have dropped significantly across Europe, but scholars find it hard to assess whether emission trading was the cause, or simply a coincidence. For instance, one study published in 2020 claims that during the years 2008 to 2016, the ETS was responsible for a mere 3.8 percent reduction in total EU emissions. That might seem modest, but it does represent one billion tons of CO_2.[47]

In retrospect, the EU's ETS constitutes a "trial and error" experiment. Its proponents see it as a tribute to the ability of a large, motley, and occasionally unruly multinational consortium to learn from mistakes and improve. Detractors emphasize its "lack of ambition." California climate experts Danny Cullenward and David Victor posit:

The central problem with nearly all real-world market instruments for cutting carbon is that they lack ambition: that is, the carbon prices and emission reductions they produce are far smaller than what societies are willing to pursue via regulatory strategies. The compromises policymakers frequent make to accommodate the interests of emitting sectors end up producing markets that reflect the lowest ambition of all covered sectors. Worse, these markets are brittle and unable to respond when conditions change.[48]

There is little disagreement that Europe's ETS got off to a bad start. Several factors undermined the system's effectiveness. A combination of miscalculation and political convenience allowed the EU's trading system to dramatically *over-allocate* emissions. This created a surplus of credits, causing carbon prices in the European market to plummet.[49] By 2013, the price of a permit to emit a ton of carbon dioxide had fallen to €2.8, a trivial sum and certainly not enough to motivate firms to invest in technology development.[50]

Moreover, many key sectors were not even required to comply with the ETS. Exemptions were granted in light of concerns about carbon leakage—the fear that industries would move their production to countries with weaker regulations, or even worse, to places that did not control CO_2 emissions at all. The upshot was that Europe's most energy-intensive sectors, such as the cement, aluminum, paper, inorganic chemicals, fertilizer, and steel industries—by design—remained unaffected. And those industries that had the misfortune of being forced to participate in the ETS soon discovered that penalties were barely a slap on the wrist.

It did not take long for the EU to recognize these flaws and make changes. In 2010, it introduced an auction whereby a portion of total allowances are sold annually, rather than allowing them to be handed out for free. By 2014, an EU-wide program replaced national allocation plans, with a full 40 percent of allowances allocated through a centralized auction. Auctioning not only solves the problem of grandfathering, but creates income: since 2017, revenues from these ETS auctions increased from €3 billion to €25 billion in 2021. These revenues are returned to the twenty-seven EU member states, with 75 percent of these funds used to address climate change.[51]

In July 2021, the European Climate Law came into force and significantly strengthened the trading system. For instance, it extends carbon pricing to new sectors—such as buildings, roads, and maritime transport—industries that had previously been exempted from the ETS. The EU's "Fit for 55" program is designed to cut emissions to 63 percent of 2005 levels by 2030. When that happens, far fewer allowances will be available, and the carbon market will become a more salient factor in a firm's decision-making. Prices are already beginning to reflect these changes. For the first time, in February 2023 EU carbon prices reached a record high of €100 per ton of CO_2.[52] It took twenty years, but the system may be starting to deliver.

China's Tradable Permit System

In 2021, China began to implement a carbon-emissions trading program. The implications of this development are potentially enormous. As the world's largest emitter of greenhouse gases, the dimensions of the new Chinese initiative are almost unimaginable: 2,225 power operators were the first group required to purchase emissions permits, with cement, petrochemicals, chemical, aluminum, and steel industries slated to join the system during a later phase.[53] Once it becomes fully operational, the Chinese program will more than double the number of polluters, worldwide, that need to a pay a price for the carbon they release.[54] The Chinese program was only implemented after policymakers learned a thing or two from a pilot program that ran in five of the country's major cities: Beijing, Chongqing, Shanghai, Shenzhen, and Tianjin—along with two provinces—Guangdong and Hubei.[55]

The pilot programs experimented with different forms of trading allowances and permits. Some used a "tradable performance standard," which was eventually adopted at the national level. Under this system, facilities must stay below a standard ratio of CO_2 emissions relative to production output. This means that there is no absolute cap on emissions in terms of tonnage, as in the European system. Nonetheless, as in all cap-and-trade programs, facilities buy and sell emissions allowances, allowing firms to reduce emissions at the lowest cost while encouraging innovation.[56]

The program is still in its incipient stages, so it is too early to judge its efficacy. Nonetheless, a team of Stanford University, Peking University, and Tsinghua University researchers, headed by Professor Lawrence Goulder, conducted a general assessment and found that the program's environmental benefits far exceeded anticipated costs: "by a factor of five if only the climate-related benefits are considered, and by a significantly higher factor if health benefits from reduced emissions of local pollutants are also considered."[57]

At the same time, the Chinese system was deemed relatively lenient, and the researchers calculated that the environmental benefits of lowering the allowances would exceed the additional expenses. In any case, the costs per ton of carbon are expected to change significantly over time. Under the new Chinese system, allowances initially were allocated for free, but China's policymakers are considering an auction requirement, much like the veteran EU system uses today. This is expected to lower the economic costs of tradable permit standards by as much as 40 percent.

Environmentalists remain suspicious of China's true intentions to move to net zero. There is no shortage of reasons to be circumspect. The ongoing approval of domestic coal-fired plants—and enthusiastic exportation of such polluting facilities—send a very dubious message.[58] Canadian environmental leader David Suzuki reminds his readers that China's reporting on environmental matters has traditionally been less than trustworthy.[59] Elizabeth Economy, director of Asia studies at the US Council of Foreign Relations, has a similar concern: cap-and-trade programs only work when they are transparent, and China has never been a paragon of candidness.[60] And yet, if local plants begin to trade emissions and if reliable monitoring shows lower carbon footprints, China's new program should be welcomed.

Emissions Trading and Innovation

Looking to the future, a key question is: Can emissions trading systems galvanize climate tech? Innovation, especially climate innovation, does not happen overnight, and neither do the effects of incentives created by emissions trading programs. Since allowances are slowly ratcheted down over time, it can take a while for companies to feel the pressure to invest in new technology development. That explains why most studies of cap-and-trade programs evaluate the veteran European system, where results should already be apparent.

The EU always intended for its carbon market to help drive the development of new technologies. For instance, copious allowances were set aside as part of a "New Entrants Reserve," with the goal of funding renewable energy technologies on a commercial scale, as well as carbon capture and storage demonstration projects that were included in the EU's "NER 300" program.[61] Unfortunately, it is not clear that these objectives were attained.

In fact, qualitative studies point to a rather anemic (if any) effect on R&D. One major assessment, involving interviews with 770 managers from firms in six participating countries, found *no* difference in climate tech development between companies that were included in the emissions trading scheme and those that were not. When companies did invest in reducing greenhouse-gas emissions, the researchers characterized their efforts as "process innovation" rather than "product innovation" per se. It is also noteworthy that when companies chose to spend corporate resources on carbon innovation, they were motivated by concerns about *future allocations*. Moreover, when firms did *not* receive free allowances, they were more inclined to pursue a low-carbon innovation strategy.[62]

Other studies, in Sweden, have found similarly underwhelming results. One set of researchers attributed the lack of initiative to high price volatility, along with the time required for corporations to understand how a carbon market functions. A later study that interviewed 706 Swedish firms pointed to low carbon prices as the reason why companies were not investing in new technologies.[63]

Quantitative research paints a somewhat more clement picture. In one of the first systematic evaluations of the EU cap-and-trade system, results seemed generally positive. The researchers found that when the EU trading system first got underway, between 2005 and 2009, there were in fact 36 percent *more* requests for patents involving low-carbon technologies registered than in analogous, non-ETS countries that lacked comparable trading policies. No comparable "indirect" increase in patenting activity was found among firms that were exempted from the trading system. Nonetheless, the researchers ultimately concluded that any surge in low-carbon patents could probably just be attributed to the rising cost of energy, a likely outcome of emissions trading dynamics.

The story in China is the same: pilot programs that preceded the national Chinese cap-and-trade initiative in thirty Chinese provinces were also evaluated. While the researchers found that cap-and-trade programs did motivate modest innovation, it was deemed "low-quality."[64] Other case studies that focused on a range of industries, from the Italian paper and cement industries to the German electricity sector, reached the same conclusions.[65]

Despite these disappointing assessments, the EU trading program has not given up on climate tech. The mandate of the EU's new Innovation Fund is to demonstrate breakthrough technologies—with funding corresponding to 450 million emission allowances. A second program, the Modernization Fund, is designed to support investments that can modernize the power sector and energy systems, enhance efficiency, and provide special support to the ten lower-income EU member countries.[66]

The Political Limits and Potential of Economic Incentives

For decades, many economists have seen carbon taxes and emissions trading programs as the most promising way for governments to reduce greenhouse-gas emissions and stimulate climate tech innovation. But reality is sticky. *The actual impact of both carbon taxes and emissions trading has largely been disappointing.*

The problem is not just that the devil is in the details. Rather, policies are only as formidable as the political will that backs them. In countries across the world, governments have been too afraid of voter backlash and the ire of influential industries to create programs with teeth. As a result, taxes were set too low; emission allowances allocated too generously; exemptions to entire industrial sectors granted too freely.

Danny Cullenward and David Victor devote most of their insightful book *Making Climate Policy Work* to pointing out the fundamental flaws that continue to characterize market-based climate-mitigation systems around the world. They summarize the dynamics like this:

> Many pollution markets exist, but nearly all are smokescreens that create the impression that market forces are cutting emissions when, in fact, other policies are doing most of the real work of decarbonization. Almost everywhere that market systems are in place, they operate at prices that are so low as to have little impact on key decisions such as whether to invest in or deploy new technologies. After thirty years of policy attention to climate change and twenty years of active efforts to design market systems, jurisdictions with reasonably ambitious carbon prices—say $40 per ton of CO_2-equivalent account for less than 1 percent of global emissions. Those with carbon prices approaching $100 per ton of CO_2 equivalent—a strong signal more consistent with the level of effort the best new science suggests is needed for deep decarbonization—are an even tinier sliver of the global picture.[67]

They conclude that the problem with markets is fundamentally political.

While it is always easy to blame the politicians, there is ample evidence to suggest that the fears of elected officials are not irrational. In 2018, France introduced a carbon tax that translated into higher gasoline prices. Protests against the tax turned violent, causing four deaths, injuring 250 people, and causing millions of dollars in damage.[68] Part of the outcry may have been class-based, as several rarefied commodities—like aviation fuel—were exempted from the tax.[69] But this collective unwillingness to pay for carbon pollution took place in a country that only three years earlier had taken great pride in hosting a successful UN climate summit that resulted in the vaunted Paris Agreement. President Emmanuel Macron and his government were quick to cancel the levy.

Five years earlier, carbon taxes became the focus of the Australian election campaign when opponents tapped into the public's distrust of government and launched a vicious attack against the policy.[70] "Axe the Tax!" cried Liberal Party chair, Tony Abbott, and many Australians listened.[71] Demonization of carbon taxes and the UNFCCC is considered a central reason for the 2013 collapse of Australia's Labor government and its replacement by a Conservative coalition. The new government did not hesitate to implement its election promise, canceling the carbon tax immediately upon assuming power in 2014.[72] In the United States, between 2003 and 2007 three separate attempts by two popular senators, Republican John McCain and Democrat Joe Lieberman, to introduce bipartisan legislation containing a mandatory cap-and-trade system for greenhouse gases failed spectacularly.[73]

Subsequent research shows that resistance to pricing carbon is quite widespread and can be characterized fairly predictably. The higher taxes are, and the faster they are introduced—the greater voter resistance.[74] There are many other factors that tend to affect individual opposition, including distrust of government, perceived fairness of the tax, educational level, and most of all, how much out-of-pocket expenses the tax imposes on the public. The upshot is that even climate-concerned politicians are unwilling, or unable, to adopt economic incentive policies that produce sufficient pressure to change behavior. An entire literature of evaluation research is fairly consistent, concluding that the outcomes of incentive programs thus far have been modest.

All the same, on the rare occasion when politicians do step up to the plate, the calculus can change. Recently, the New Zealand government moved to plug the holes in its carbon market. The historic lenience toward agricultural emissions did not escape the attention of the country's Labour government, which takes climate policy very seriously. New Zealand was among the first countries to make a commitment to net-zero emissions by 2050. And so it was that in October 2022, New Zealand proposed the world's first carbon tax on cow emissions.

Roughly half of New Zealand's carbon footprint comes from farming, three-quarters of which involves methane from livestock.[75] With 6 million cows, 25 million sheep . . . and 5 million people, this is hardly surprising. The country's idiosyncratic zoological portfolio has always characterized Kiwi emissions. In 2003, New Zealand's government, traditionally a paragon of integrity on the international stage, moved to impose a carbon tax on its agricultural operations. But when an angry mob of a thousand farmers charged the parliament in Wellington, it backed down.[76]

Twenty years later, national awareness and commitment to addressing the climate crisis had matured. The actual rate for New Zealand's new livestock tax will depend on the dimensions of farms, herd size, fertilizer types, and concrete steps taken to reduce emissions from farm animals. But the goal is clear: cutting methane emissions by 47 percent by the year 2050.[77] This round, protests by farmers were more muted and the parliament building was not attacked by a squadron of tractors. Nonetheless, agriculture advocates expressed legitimate concern that when implemented, the expanded carbon tax would force many farmers to reduce the size of their herds, threatening their profits. Then Prime Minister Jacinda Ardern did not blink, explaining to farmers that with the right framing, they should be able to recover the costs through the higher prices they can charge for climate-friendly products.[78]

It is quite plausible that a carbon tax, specifically targeting farm animals, will lead to innovation in the agricultural sector. Many promising ideas are presently being developed to reduce methane emissions, including Australian *Rumin8*, whose red seaweed and range plant supplements have reportedly cut methane output by 80 percent.[79] (The company is promising enough to have received a $12-million investment by Bill Gates and his Breakthrough Energy Ventures.)[80] Many of the best minds in animal husbandry and physiology departments at universities around the world are already reporting meaningful progress in reducing emissions from livestock.

This suggests that carbon taxes and emissions trading should not be characterized as failed public policies that need to be abandoned. Any pathway to net zero will require creative solutions beyond decarbonizing electricity and industry. Such breakthroughs are unlikely to happen without monetizing carbon at the global level.[81] If enough countries set the price high enough, leakages can be prevented. This would change the economic calculus in high-emitting industries, spawning new inventions and innovative technological alternatives. At the same time, strategically, *subsidies* for low-emission technologies probably offer a more effective route to decarbonization, engendering far less public resistance than carbon taxes or emissions trading schemes. This insight leads to the more encouraging story told in chapter 5.

CHAPTER 5

Incentives for Innovation

As president of the United States, Jimmy Carter was troubled by the oil crises of 1973 and 1979, with the resulting images on the nightly news of long lines of drivers waiting to pump gasoline into their cars. Carter recognized that global fossil fuel supplies were limited and believed that shortages were certain to grow worse. As president, he established a Department of Energy and soberly declared America's energy crisis to be the "moral equivalent of war."[1] Action was needed.

Wearing a cardigan sweater as he spoke to the nation in front of a fireplace in the White House, the president called energy conservation "the cornerstone of our policy." Photovoltaic solar electricity at the time was still largely a marginal, undeveloped technology. But Carter was an engineer and far more technically literate than your typical politician. He appealed to Congress to support research and development in solar energy. In April 1977, the president formally proposed a National Energy Program, which for the first time subsidized renewable energy through tax credits—after decades of massive subsidies generously doled out to oil and gas interests.[2]

Under Carter's Energy Program, citizens could deduct 40 percent from the first $1,000 they spent on solar equipment, and 25 percent from subsequent outlays, up to a $6,400 ceiling. Businesses would enjoy a flat 10 percent tax credit for energy-saving expenses. Moreover, Carter set up a Federal Energy

Management Program that included a three-year initiative to install solar equipment in government buildings as a demonstration of the country's support and confidence in solar electricity. States were called upon to lend a hand by exempting solar installations from any property taxes as well as promoting "solar education" to consumers. In 1979, the president put his money where his mouth was, installing solar collectors for a water heating system on the White House roof.[3] Carter's policies made a difference. A series of studies identify the late 1970s as the start of a steady drop in the price of solar power, albeit at the time, only passive solar water heating systems were even remotely cost-effective.[4]

It would not take long for Carter to lose the 1980 election to Ronald Reagan. The new president's cheerful "morning in America" disposition was widely seen as an antidote to Carter's "gloomy" focus on natural resource constraints. By 1986, Reagan had rolled back Carter's renewable tax credits,[5] along with unceremoniously removing the solar system from the White House roof. (Its fate constitutes a fascinating renewable anecdote: After five years of collecting dust in a Virginia warehouse, an enterprising administrator from Unity College in Maine drove a school bus 700 miles south and liberated the equipment.[6] One of the thirty-two orphan panels ended up in the Smithsonian Museum, one in Carter's presidential library, and one in China.[7] In 2013, twenty-six years after Carter's dramatic gesture, renewables had the last word: President Barack Obama, reclaimed the White House for solar electricity. This time a PV system was mounted on the White House roof.)[8]

Reagan could physically remove Carter's panels. But he could not stop the reverberations of the bold incentive policy that continued to resound across America, soon reaching Europe. Countries like Sweden, France, and the Netherlands began to offer tax exemptions for renewable energy systems.[9] In 2005, after the wheels of politics turned yet again, solar tax credits under the Clinton administration jumped from 10 percent to 30 percent, with an immediate effect on the solar market. Within three years, America's solar energy capacity tripled.[10]

The 30 percent investment tax credits proved resilient, surviving a range of both Democratic and Republican administrations. Most recently, Carter's original tax credits were extended again: the present iteration appearing in the US Consolidated Appropriations Act of 2021. The Act also grants a 30 percent credit rate to offshore wind systems, as well as a somewhat lower deduction for heat pumps, small wind, waste energy recovery, and geothermal systems.[11] Businesses also became eligible for a "production tax credit." This credit is calculated based on the actual electricity generated by solar systems (measured in kilowatt-hours). Commercial enterprises using other qualifying technologies can also continue to deduct 30 percent during the first ten years of a system's operation, adjusted annually for inflation.[12]

Carter's renewable tax exemptions and their later manifestations were important initial steps toward government support of renewable energy. After all, tax credits already had a long, illustrious history of shaping public behavior.[13] Long before there was a climate tech industry, tax breaks in America were used to subsidize college education, encourage employee pensions, support house ownership, and finance countless charities.[14] In the same way that people tend to overvalue things that are free,[15] tax exemptions leave many people with a feeling that they've hit the jackpot, enticing them—or their company—to do things that ordinarily they would have never considered.

Rebates on climate technologies, feed-in tariffs for clean energy, and carbon contracts for difference, all share the appeal underlying tax credits. Each has proven highly effective in influencing decisions to adopt climate technologies. Change did not always happen overnight. Indeed, it took nearly thirty years for American tax credits to become a meaningful driver of renewables, dramatically expanding solar systems on rooftops across the United States. That only happened after a serial entrepreneur from California, David Arfin, had an epiphany.

SolarCity Discovers Tax Exemptions

Having grown up in Silicon Valley and attended Stanford Business school, Arfin had stellar entrepreneurial bona fides and environmental impulses. But he had yet to discover his passion for climate tech. In 2006, he watched the Academy Award–winning documentary by Al Gore, *An Inconvenient Truth*. It changed everything for him. Arfin decided that he had to do something to get solar energy systems on people's roofs. The problem was that the price of renewables was still prohibitively high. Looking back, he shares his thinking at the time: "I said to myself—it will take me thirty-three years to get back my investment in a rooftop PV solar system on my home. That means that installing photovoltaics is essentially a donation. I thought, if I'm going to give away $40,000 for the environment, I might as well buy up ecological hotspots or conserve acreage in Costa Rica."[16]

So he started looking at the US tax code for inspiration and found that while businesses enjoyed healthy tax breaks for installing solar systems, the deduction for individual homeowners, on the other hand, was capped at $2,000, even though the price of a system could run $40,000 or more. Similarly, businesses could declare tax exemptions on the depreciation of the system over time;

homeowners were not eligible. But then California governor Arnold Schwarzenegger offered a significant rebate for renewables that covered about third of the initial costs. Looking back, Arfin recalls, "With this sort of government support, a business could completely derisk this new asset class. So that's actually what we did: we created a new asset class."[17] In short, while it made no sense economically for a household to purchase a rooftop renewable energy system, for a corporation, it was a bonanza.

Arfin realized that if businesses leased solar systems to individuals and passed along some of the tax savings, it would be a win–win for both. When he shared his insight with Peter and Lyndon Rive (cousins of Tesla founder Elon Musk), who had just started a new solar company, they immediately grasped that Arfin was onto something big and got out of his way. Literally going door to door, the company's sales teams fanned out to offer California residents an opportunity to be part of SolarCity's leased systems. It was an easy sell: new solar customers did not have to shell out any upfront costs at all. They simply agreed to pay the solar company a *reduced* rate for the electricity generated by the panels on their roofs for twenty years.

Given California's high electricity rates (about a third above the national average)[18], in the long-run, tax exemptions produced a profitable bottom line. But the company still needed a deep pocket to finance the solar systems up front. After the 2008 home mortgage crisis, this was challenging, because the financing agreement for solar systems essentially created a second mortgage at a time when defaults were common. Eventually, the investment bank Morgan Stanley agreed to step in and provide the missing piece.

Arfin left SolarCity in 2011, but by then the tax credit–driven business model had already become the predominant strategy for increasing solar in the United States. Although SolarCity initially purchased American-made solar panels, it did not take long for Chinese companies to offer even lower prices on panels, and they soon dominated the market. With its early start and financing muscle, SolarCity promptly emerged as the country's largest residential solar installer, signing up customers in dozens of states before being purchased by Tesla in 2016. By then it had acquired a few billion dollars in debt to cover the costs of its system, but its income stream to cover payments was steady. (Arfin explains that even the most unreliable citizens usually pay two of their bills: electricity and cable.)

Fueled by a combination of government tax exemptions and descending prices, America's solar energy industry went from an extremely modest niche for a tiny elite, to a multi-billion-dollar industry for the masses. (Consider the math: 2 million households deploying $30,000 rooftop PV solar systems comes to $60 billion.) At its height, SolarCity enjoyed a third of the market share. With demand for photovoltaics growing exponentially, American enthusiasm for renewable energy pushed solar systems even further along the learning curve, which meant ever lower prices.

Net-Metering and Paying for Grid Infrastructure

This democratization of solar was helped along by a "net-metering" policy, already adopted in many states across the United States. Beginning with Arizona, Idaho, Massachusetts, and Minnesota back in the 1980s, homeowners were no longer required to actually use the power they generated from their solar panels. Rather, everything was fed back to the grid and individuals received maximum monetary credits for the kilowatt-hours they produced.

This flexible billing mechanism means that a consumer can receive electricity from the grid at night in return for the electricity she sends to the grid in the afternoon. It also means that in the dead of winter, a family can take advantage of the electricity that its solar system generated when summer sun was abundant. Given the inherent intermittency of sunlight, the ability to receive credits is critical for household solar and wind systems to be profitable, especially when they are installed without storage systems. Net-metering meant that SolarCity's customers received full retail price credit for every kilowatt-hour they sent to the grid. During the solar boom in California, this could reach 45 cents per kilowatt-hour.[19]

When net-metering is adopted, solar homeowners not only save on electricity bills, but they can also repay initial investments in renewables more quickly. The United States is an excellent laboratory for demonstrating how net-metering can increase renewable energy use. As several states never adopted the policy, comparisons are easy. Take, for example, a household's "payback time" for renewables. According to estimates by the solar industry, it would take between four and five years to pay for a solar power system in New Jersey, given its full net-metering policy. This contrasts to twelve years for systems in South Dakota, which refrains from any form of net-metering.[20] Europe was slower to adopt the policy, inter alia, due to hesitancy on the part of the United Kingdom, which struggled to find a way to refund the associated value-added sales tax.

Electric utilities traditionally opposed net-metering policies because they undermine the business model financing electricity grids; money going back to customers means it's not going to the utilities that made the initial investment in infrastructure. Net-metering is also criticized because it exacerbates income disparities: people who own renewable systems pay utilities far less in monthly fees. As a result, the cost of maintaining and expanding the grid falls on those who do *not* generate electricity. Wealthier households tend to be early adopters of solar or wind systems. Hence the policy was considered to be unintentionally regressive. Utilities are forced to raise rates; poorer families end up subsidizing the solar transition; and affluent households profit handsomely.[21]

Still, when net-metering was first introduced, many felt the overall societal benefits justified these disadvantages. The "pros" include democratizing and de-centralizing electricity production, along with a reduced reliance on fossil fuel or coal-fired power plants, even in the face of mounting electricity demand.[22] As more homeowners connect electric vehicle charging systems to their solar systems, the overall environmental and economic benefits increase. Net-metering also allows policy makers and utilities to consider a richer variety of "non-wires alternatives"; in other words, investments in energy systems that do not rely on building more transmission lines or new power plants because demand has shifted and there is less "transmission congestion" on the grid.[23] In Australia, for instance, before a new power plant is built, the costs and timetable need to be published. Any non-wires alternative that meets the same metrics is adopted.[24]

Over time, a growing number of policymakers around the world came to agree with utilities that net-metering was unfair and economically unsustainable. Countries began to adjust their policies and electricity rates accordingly, setting the price of solar credits *below* retail electricity rates, or even levying a tax on solar companies. In 2022, for example, notwithstanding the consternation of environmental advocates, the UK treasury announced a 45 percent levy on the "extraordinary returns" of low-carbon electricity.[25] At the same time, the California Public Utilities Commission was set to impose a $50 monthly surcharge for solar-equipped customers to help cover the costs of maintaining electricity grids. The proposal faced heavy political pressure from irate solar users as well as senior California politicians (including present and past governors: Gavin Newsom and Arnold Schwarzenegger). The commissioners eventually caved in. Instead, the Commission opted for a path of least resistance by cutting the compensation paid to *new* customers exporting solar power to the grid by 75 percent below existing levels.[26] Arizona also significantly reduced the "export rates" solar customers receive, raising the "fixed charges" component of these consumers' monthly electricity bills that covers infrastructure costs—especially when electricity is utilized during peak hours.[27]

Given the urgent need to decarbonize, any retreat from net-metering policies and prioritizing non-wires alternatives appears to be premature and ill-advised. If underprivileged individuals are adversely affected by energy policies, there are a range of social welfare policy tools designed for addressing inequities. For the foreseeable future, energy policies must focus on sustainability. That's because, notwithstanding the drop in the price of renewables, electricity still remains the greatest source of greenhouse gas released globally. And unfortunately, emissions associated with electricity generation continue to rise.[28] As of 2023, a mere nine countries have solar energy penetration rates over 10 percent of total electricity supply, with only Spain, Greece, and Chile above 17 percent.[29] Slowing the energy transition because a few early adopters of renewables save a little money makes no sense. Electricity's heavy carbon footprint simply cannot wait.

As a founder of industry giant Solar Edge, Lior Handelsman has followed renewable energy policies in over 100 countries around the world and has seen the consequences of phasing out net-metering. His conclusion is that it may be perfectly logical for policy makers—but it can also be detrimental to solar energy companies, whose profit margins are already razor-thin: "You can't pull the rug out from the solar industry and expect it to continue to operate smoothly." Handelsman offers sage advice about how the solar industry should respond to such sudden policy changes:

> Companies start to publicly protest—and that's probably the worst thing they can do as an industry. Here's why: the government doesn't care; and generally, the public doesn't care, either. What it does hear, though, is that solar energy is no longer profitable. And that's not true. Solar electricity is still profitable. Just less so for the solar companies, who can pass on the additional expense to consumers. Ultimately, the sales damage will be ten times worse than the financial damage. The only place in the world where this doesn't hold is the US. There you go out and hire a lobbyist![30]

After the passage of the Inflation Reduction Act, the American solar industry does not need to do a whole lot more lobbying.

The IRA: Tax Credits on Steroids

All over the world, tax exemptions have proven to be an effective, politically feasible way to scale renewables. This surely wasn't lost on the architects of Joe

Biden's climate strategy. Alongside its focus on encouraging domestic production of low-carbon technologies, the Inflation Reduction Act is primarily aimed
at boosting existing climate technologies through tax credits to both households
and the commercial sector. Indeed, nearly three-quarters of the Inflation Reduction Act's clean-energy investment is delivered via tax incentives, putting the US
Department of the Treasury at the forefront of this landmark legislation.[31]

When the bill was signed into law on August 16, 2022, the Treasury Department did not even wait twenty-four hours before issuing guidance to help consumers claim an IRA credit of up to $7,500 after buying an electric vehicle.[32] But
it was not an automatic handout. To get the full credit, the law requires that 60
percent of the value of battery components be produced or assembled in North
America and that 50 percent of the value of critical materials be domestically
sourced or come from a country with whom the United States has a free-trade
agreement. The market response was immediate: in 2019, there were only two
plants manufacturing batteries in the United States; a year after the law was
enacted, some thirty battery factories are either planned, under construction, or
operational.[33] And EVs are just the tip of the iceberg.

Additionally, the IRA offers solar developers and US manufacturers of solar
components more tax advantages than you can shake a stick at. Ethan Zindler,
head of Bloomberg New Energy Finance, has overseen the company's monitoring and analysis of clean-energy trends for years. He confirms that the incentives
for solar production and installation are nothing less than a game changer: locally
produced modules, cells, wafers, polysilicon, and inverters are all subsidized.
Zindler jokes: "Each segment of the value chain has been shown some love."[34]

One bonus introduced by the IRA is the ability of solar developers to take
advantage of production tax credits (PTCs) in lieu of investment tax credits
(ITCs) if they are more lucrative. Production tax credits have the advantage
of paying out based on the amount of energy produced, while investment tax
credits are only granted for actual dollars spent. Zindler explains the new dynamics: "If you are developing a large project in a very sunny part of the US, the
production tax credit may offer a substantially more generous benefit for you
as a developer." Assuming a continued drop in PV prices, this preference may
become a no-brainer.

Moreover, assuming that developers are paying "prevailing wages"[35] (the
average pay received by workers for solar installation), the IRA offers even more
exemptions to taxpayers: A 10 percent credit can be attained if the solar system
is domestically produced. If a renewable facility is located in a low-income area
or community hit hard by the energy transition, credits can be ramped up by an

additional 10 percent. Even before the law was enacted, solar energy was already the cheapest way to produce electricity. For savvy developers, the IRA essentially cuts the price in half.[36]

What will be the effect of the policy on solar production? Princeton University professor Jesse Jenkins offered the first projections as part of his REPEAT (Rapid Energy Policy Evaluation and Analysis Toolkit) project. In 2020, 10 gigawatts of solar capacity was installed in the United States; by 2023, installation will exceed 30 gigawatts (GW); and by 2030, Jenkins estimates a steady increase to 100 gigawatts or even beyond.[37] Climate tech maven Shayle Kann sees that amount as nothing less than "mind-boggling," saying, "Thanks to the IRA, we have long-time certainty on the tax credits, along with a bevy of other boosters for solar components and products to ensure that the cheapest kilowatt around is going to be a solar kilowatt hour."[38]

Under the IRA, in 2023 and 2024 only solar and wind technologies, along with associated storage projects, are eligible for credits. But by 2025, the policy adopts a technology-neutral orientation. Tax credits become available for disparate climate tech alternatives, even as credits for wind, solar, geothermal, and nuclear, of course, continue.[39] Critically, the law creates a ten-year "runway" for these new tax exemptions, providing investors and manufacturers with the certainty they need for investing in low-carbon alternatives. Projects involving energy storage systems (e.g., stationary batteries; stationary hydrogen fuel cells; hydrogen storage vessels; microturbines for combined heat and power systems; pumps and turbines for pumped hydropower storage systems; and specialized components of any such equipment) are also fully eligible.[40]

But the menu of IRA tax incentive programs expands far beyond renewable energy. The options are so numerous that when the White House published an online "Inflation Reduction Act Guidebook," the explanatory Excel file contained 113 lines, all describing different programs run by myriad federal departments and agencies.[41] Any one of them could be a game changer. For instance, REPEAT modeling projects that with IRA incentives, by 2030 electric cars should make up 41 percent of new light-duty vehicle sales (and possibly more than half)—about 25 percent more than sales without the government support.[42]

Tax exemptions under the IRA are generally linked to production volume. This explains why estimates of the actual credits that will ultimately be granted under the law vary so dramatically. For instance, for every kilogram of green hydrogen produced, the government will allow a $3 deduction.[43] Energy expert Robinson Meyer calculates that IRA subsidies essentially mean that the cost of producing green hydrogen will be fully covered by the government, allowing it to be price-competitive with natural gas. The Congressional Budget Office assumes that only $5 billion in tax credits will be used by domestic electrolyzers for hydrogen production. But some experts in the Department of Energy believe that if there is a significant build-out of the American hydrogen infrastructure, it could be as much as $100 billion.[44] The same holds true for carbon capture and storage.[45]

There is also a special $10-billion investment tax credit for clean-energy manufacturing and another $6 billion to support heavy manufacturing, such as steel and cement. And if all this weren't enough, American clean energy will receive an additional boost from a $250-billion expansion of the Department of Energy's loan program.

Economic models were optimistic about the IRA's ability to stimulate private investment in climate tech. It took only a year for empirical effects to be measured. The impact of tax credits actually exceeded initial expectations. The Clean Investment Monitor, a database set up by the Rhodium Group and MIT looked at millions of facilities across the United States and reported on climate tech progress in autumn 2023, twelve months after the IRA's adoption. Overall investment that year in American clean-energy climate tech manufacturing reached $213 billion, up 37 percent from the previous year—and 165 percent higher than levels five years prior. Most dramatic was the expansion of actual investment in manufacturing as part of the EV value chains—including critical minerals, batteries, EV assembly, and charging equipment: up 93 percent over a two-year period in the United States.[46] These dramatic figures reflect the snowball effect created by ambitious government subsidies for decarbonization, and they validate the claims of an American cleantech industrial renaissance.

Before the passage of the IRA, the 30 percent projected drop in American greenhouse-gas emissions by 2030 fell short of the United States' international 52 percent commitment.[47] With the anticipated adoption of climate technologies, driven primarily by statutory tax credits, it is now likely to reach 40 percent and beyond.[48]

Feed-in Tariffs

The International Energy Agency makes a distinction between tax credits for renewables, which are seen as subsidies, and feed-in tariffs, which instead provide a guaranteed, above-market price for renewable electricity producers. While credits reduce upfront costs, feed-in tariffs are simply a way to ensure a handsome return on investment.[49] Regardless of their formal classification, feed-in tariffs provide powerful incentives for individual and corporate renewable energy producers to generate electricity and then sell the surplus back to the grid.

These tariffs deserve credit for plummeting solar prices at the turn of the twenty-first century—one of the few bright spots in a generally underwhelming period in climate mitigation history. At last count, some 107 countries, including Spain, Japan, Canada, and South Africa, had adopted feed-in tariffs at some point to encourage the expansion of solar energy.[50] Yet no nation has had more success with the policy than Germany. It was officially adopted in 2000 as part of Germany's Renewable Energy Sources Act, which galvanized *a dramatic transition to a low-carbon, environmentally sound, reliable, and affordable supply of electricity.* Providing guaranteed payments to renewable energy producers, Germany's feed-in-tariffs achieved results—with unusual efficacy.

The German feed-in tariff story actually begins in the 1980s, when citizens were growing more and more worried about the dangers of nuclear energy. Germany's first nuclear reactor went online in 1961. Many more would follow. Within twenty-five years, some 20 percent of the country's electricity production came from nuclear plants. But 1986 was also the year of the meltdown and explosion at Chernobyl, the worst nuclear accident in history.[51] With more than 350,000 people from 213 towns forced to relocate outside the contamination zone in Ukraine,[52] Germany's nascent Green Party—then only six years old—demanded the suspension of all nuclear power generation in the country.[53]

A decade later, in the 1998 election campaign, the Greens ran on a platform of bringing solar power to 100,000 rooftops. When the votes were counted, they had won forty-seven seats and joined a ruling coalition that quickly became known as the "red-green" government. The party immediately became the driving force behind the country's renewable energy strategy, and champions of new feed-in tariff legislation. Previously, in the mid-1980s, Denmark had dabbled in feed-in tariffs with moderate success.[54] But fueled by the Greens' political passion, Germany would soon take the policy to a new level.

In 1999, the party kept its campaign promise and Germany launched an initiative to install 100,000 solar PV systems on the roofs of business and private homes across the country by 2003. To jumpstart the program, property owners could apply for a loan of up to €50,000 that would cover the cost of purchasing and installing a solar PV system. The low-interest, 3–4 percent loans were guaranteed by a government-owned development bank and had a repayment period of ten years.[55] A year later, feed-in tariffs were introduced. They provided a reliable income stream to cover loan recipients' monthly expenditure.

Under the policy, if a grid connection was considered "necessary and fea-sible," families and businesses that installed solar photovoltaic systems on their roofs were entitled to sell the electricity they generated back to the grid at a hand-some, fixed rate . . . for a period of twenty years. The policy objective was not just to promote renewables, but also to democratize the country's electricity system, which was dominated by large, centralized utilities that were largely reliant on burning coal and running nuclear reactors. The new strategy was essentially an intervention on behalf of scores of new, dispersed potential generators: either small and mid-sized utilities or families with a roof.

The initial rate paid to early adopters was positively lucrative: on average, 51 euro cents per kilowatt-hour. At the time, this covered 90 percent of the average costs of solar system paybacks for private customers. In other words, the govern-ment loan for installing the solar system essentially was transformed into a grant. The Greens saw the tariff as both an effective path away from dependence on nuclear power and an opportunity to challenge the monopoly held by large Ger-man utilities. They were right on both counts. But the tariff was not just a Green Party pet project. It enjoyed support from allies in the Christian Social Union Party as well as the Social Democratic Party, the largest member of the ruling coalition. The Social Democrats preferred to emphasize the merits of introduc-ing competition to the electricity system, as well as the potential market for pho-tovoltaic electricity technology worldwide once the cost of PV panels dropped.[56]

Results on the ground (and on the roofs) exceeded even the Greens' origi-nal grand vision: by the end of 2003, over 130,000 rooftop PV systems were generating over 1.1 gigawatt-hours of electricity. The program set Germany on a sustainable trajectory that has continued ever since. As energy analyst Arne Jung-johann explains, in 2022 German production from renewables reached a new high: 46 percent of electricity consumption. That means that on sunny days, more electricity is being produced during the day than is needed domestically.[57]

Just as there is no free lunch, however, there are no free photovoltaics. As renewable capacity grew, so did the amount of money paid out by German citi-zens, who footed the bill for the country's solar revolution. In 2001, Germany's overall public investment in renewables came to €6.2 billion; by 2010, the sum had grown fourfold, peaking at €27.8 billion.[58] In order to circumvent EU pro-hibitions on "unfair" government subsidies, payments were made to PV system owners by utilities. In response, the utilities hiked consumers' electricity bills. After the additional surcharge, by 2010, German households were paying over 20¢ (US) for a kWh of electricity[59] at a time when average electricity prices in the United States were roughly 8¢/ kWh.[60]

Beyond disgruntled consumers who had not joined the feed-in bonanza but were underwriting it, the tariff program had its share of institutional enemies: from the policy's inception, the German Federal Ministry for Economic Affairs worked to have the costly program rescinded. The European Commission came out with criticism of "illegal state aid" and started an investigation. Of course, the "Big Four"—the country's predominant electrical utilities—continued to resist the headache of intermittent power generation and the low return of renewable sources in favor of coal and nuclear power.

The lobbying to ameliorate the impact of the "solar bias" ultimately proved successful. Based on concerns about competitiveness, the German government exempted energy-intensive industries from having to pay the higher electricity rates charged to cover the feed-in tariff payments. As the cost of solar technology decreased, and the number of solar installations swelled, Germany decided to ratchet down tariff payments. This made sense economically. In 2012, the government introduced a system to gradually reduce the tariff rate each year, according to a formula based on the number of new solar installations. This helped prevent sudden drops in tariff rates, which could have negatively affected the solar industry. By then, Germany's emissions-free electricity revolution was well on its way. But financial considerations, grid capacity, and the lack of storage to balance power intermittency all suggested that it was time for a new approach.

In 2014, Germany moved to a reversed auction system as its central policy for stimulating new solar electricity capacity. This involves a competitive bidding process where multiple suppliers compete for renewable energy contracts. Essentially, the government sets its renewable energy targets for a given year and then invites bids from developers for projects that can produce a comparable amount of energy. The program is credited with driving down prices, as well as allowing for greater transparency.

The German reverse auction system for renewable energy has been criticized by some for being too complex and bureaucratic, which discourages small-scale solar generators from participating. The system favors large-scale projects over small ones by design, as larger projects typically are able to offer lower prices due to economies of scale. The new approach means that most of Germany's additional solar capacity comes from commercial facilities. The rate of solar electricity expansion has also become considerably slower than during the renewable "glory decade" that was spawned by the feed-in tariff sweepstakes.

Under the new system, the potential rooftop contribution of solar electricity is not entirely abandoned. A modest quota for small-scale solar installations (up to 750 kW) is set aside during auctions. But Germany's feed-in tariff boom years were over. The policy instrument had done its job, and the country remains on track to easily meet an 80 percent renewable target by 2030.

The feed-in tariff story has been the same all over the world. Denmark's unique position as a leader in wind energy generation and technology can be attributed, in part, to its feed-in tariff for wind power. Established during the 1990s, the Danish policy paid individuals and communities that generate wind energy for the electricity they produce, regardless of whether it was consumed on-site or fed into the grid. The actual price of the tariff varies according to the size and type of wind system, as well as the amount of electricity generated. Historically, the tariff was far higher than the market price of electricity. This not only created a financial incentive for individuals and communities to invest in wind energy projects; it also helped reduce the risk associated with the significant investments required for installing turbines.

Spain started its tariff program in 2008, almost a decade after Germany's launch. It is credited with renewable energy sources, rapidly making up half of the country's total electricity generation. Soon thereafter, in 2011, China began its feed-in tariff program.[61] Much like the German progression, eight years later, in 2019, the Chinese tariff was phased out when a market-based approach was approved for supporting renewable energy.[62] By then, feed-in tariffs had transformed the world's most populous country. During the course of a decade, national solar electricity installations increased by 2 million percent: from 80 megawatts of solar electricity to 204 *gigawatts*, with China soon surpassing Germany as the world's largest PV market. Indeed, by 2023, solar capacity in China was greater than the rest of the world combined.[63]

The Chinese tariff incentives were not offered universally across the country, but rather were gradually phased into select districts. The incremental implementation created a simple experiment. Renmin University scholars conducted a "difference-in-difference" analysis to compare solar adaptation in tariff zones with solar adaptation in regions where homes were not offered the bonus. The results were exceptional. The researchers concluded that by providing consumers with a tiny, 1.4-cent guaranteed payment per kilowatt-hour for selling back their electricity, 18 gigawatts of installed zero-emissions electricity capacity was added to the national grid. For once, the usually reserved econometricians did not understate the implications of their findings: "If China did not have any PV subsidies, the PV deployment market would virtually disappear."[64]

The major advances in the solar market in China, and throughout the world, can be traced back to Germany's early leadership on feed-in tariffs. Solar Edge CEO Zvi Lando is effusive in his praise of the German approach to feed-in tariffs as a catalyst for solar innovation and scaling:

Essentially, the Germans kick-started the industry. They managed the program with great consistency, notwithstanding the changes in government. Then they lowered the price gradually until it essentially disappeared. Today, the drop in German feed-in tariff means that the return for selling electricity back to the grid is so low that it simply doesn't make sense. This encourages homeowners to go to the next stage as prosumers [people who both produce and consume energy] and install battery storage.[65]

Indeed, by 2023, following the Russian invasion of the Ukraine, electricity rates were higher than ever in Germany. Households pay 40 euro cents per kWh—compared to 17 cents in the United States.[66] In fact, Germany's dramatic phasedown of tariff payments may indeed spawn a second German solar revolution. With hundreds of thousands of households deploying photovoltaic infrastructure on their roofs, there is a real incentive today *not* to buy electricity from the grid at night, but rather to invest in batteries and use the stored electricity generated during the day.

California has already formally integrated these German dynamics into its tariff strategy. The state's new "Net metering 3.0" policy transformed solar electricity's economic equation, and in so doing, it dramatically expanded the state's battery storage capacity.[67] In practice, California decided that export rates for solar energy would no longer be uniform and would fluctuate widely according to hour. On average, homeowners might receive as little as 5¢ per kWh when they sell electricity to the grid. But rates for solar systems with storage capacity, able to provide electricity between 6:00 p.m. and 8:00 p.m., when renewable energy generation tapers off and the grid is starved for electricity, will increase to $3.00/kWh![68]

Accordingly, after April 2023 net export of solar electricity only brought households a modest energy credit rather than the previous full retail value (including a transmission/distribution component).[69] As solar and storage entrepreneur David Arfin explains: "You used to get credit for retail, but now it is only for wholesale."[70]

This changes the calculus for battery storage purchases. Solar Edge general manager Amir Cohen estimates that if a homeowner has excess electricity to sell during the high demand hours, additional potential profits could reach $850 a month. Naturally, adding battery storage increases the up-front cost of a solar system. But the new rebates offered under the Inflation Reduction Act for energy efficiency retrofits offset much of the extra costs. Single-family homes are eligible to $4,000; multifamily buildings can receive as much as to $400,000—amounts that can cover roughly half of the outlay.[71]

The results of the policy shift were fast in coming. Individuals and companies with solar systems immediately understood that batteries needed to be part of their renewable package. This would allow them to generate electricity during the day and then store it until a more propitious hour. Only six months after the new rules were adopted, Tim Dade, chief operating officer of Scale Microgrid Solutions reports that basically, there are no longer "solar-only" projects being built in California: "People realized that they had to have a battery in order to optimize economic value."[72]

Ironically, the initial impulse for many of the German politicians supporting the feed-in tariff was to create favorable conditions for German solar companies, offering them a relative advantage in developing solar electricity systems for the international market. After entering the photovoltaic market in the early 2000s, Munich-based corporate giant Siemens was considered to be the world leader in solar panel production. But it did not take long before Siemens could neither compete with the performance nor the low prices of photovoltaic manufacturing in China. By 2014, the company acknowledged defeat, announcing that it would focus its renewable energy activities on wind and hydro power.[73]

Notwithstanding the German solar manufacturing collapse, Potsdam University energy policy professor Johan Lilliestam explains that, in retrospect, the German government's "solar gamble" paid off handsomely in employment. While Germany lost several thousand PV manufacturing jobs during this period, by 2011 the solar revolution had created at least 400,000 jobs in installation and maintenance of PV systems. Even today, 350,000 Germans find gainful employment in the renewables sector.[74]

CEOs of solar corporations are not the only ones who should be grateful to Germany for its vision in introducing feed-in tariffs. Anyone concerned about the climate crisis understands the need for a rapid worldwide transition to renewables. It is impossible to overstate the importance of the precipitous drop in solar panel prices to meaningful progress in decarbonization. In 2000, when the German feed-in tariff began triggering the proliferation of photovoltaic roofs, the median installed price of a residential solar system in the United States was $12.87 per watt of energy. By 2023, depending on the state, it was between $1.00 and $1.50[75]—a drop of over 1,000 percent—during a 23-year period when inflation has been roughly 79 percent![76]

The benefits of the "induced innovation" created by a strategic technology support program at the turn of the twenty-first century extend far beyond the life of the Germany subsidy and make PV systems affordable for people all over the world today. According to a 2018 Harvard University study, "Without accounting for induced innovation, subsidies increased global solar adoption 49 percent over the period 2010–2015, leading to over $15 billion in external social benefits. Accounting for induced innovation increases the external benefits by at least 22 percent."[77]

In a sense, the German people gave a gift to the world: inexpensive photovoltaic electricity. Lower prices are particularly critical to developing nations, which can leapfrog over economic development fueled by coal and natural gas to renewables—a move that will be essential if the world is to achieve net-zero emissions. Feed-in tariffs, an expression of Germany's climate commitment twenty years ago, make this possible today.

Subsidies Down Under

Roughly 20 percent of global greenhouse-gas emissions comes from heating and cooling buildings. Developing and disseminating technologies to dramatically cut the residential carbon footprint constitutes low-hanging fruit. In fact, of the many solutions that exist, the heat pump is among the most cost-effective.

The reason that heat pumps are more energy-efficient than conventional air conditioners or furnaces is that heat pumps don't actually create heat. Rather, the pump simply moves warm air, which requires far less energy than the electricity for cooling that air-conditioning consumes. The amount of energy that heat pumps save depends on local climates and actual home conditions, but they are generally considered to be 30–50 percent more energy-efficient than present alternatives.[78]

This savings comes at a price: heat pumps can cost as much as $7,500 up front, which constitutes a significant deterrent for many households, notwithstanding the long-term savings they ultimately will enjoy in their electricity or fuel bills. Some utilities offer modest rebates to encourage heat pump installation. An ambitious incentive program, on a scale comparable to the glory days of the solar tariffs, could defer this expense—and accelerate adoption of other low-carbon, economically superior technologies. That is precisely what Australia has started to do. The 2022 US Inflation Reduction Act embraced a similar approach, offering a $2,000 tax credit for installation. Poland, which now subsidizes heat pumps at a level of up to 30 percent, saw installation increase by 120 percent in 2022.[79]

Australia's renewable energy policies started slowly. Then a decision to phase out an aging fleet of coal-fired plants, along with conscientious policies to decarbonize electricity, led to extraordinary results that far exceed the country's initial declarations. Incentives went into high gear, and by 2023 the percentage of electricity supplied by renewables to Australian grids was around 26 percent.[80] In many Australian states, the 2050 net-zero future is already here. For instance, in 2022, the hydropower-endowed state of Tasmania reported renewable energy penetration of 99 percent, while South Australia went from below 1 percent of renewably sourced electricity to 75 percent in less than seventeen years. (Wind turbines generate a full 44 percent of present supply.) Today, South Australia exports clean electricity to its neighbors.[81]

Subsidies have played a key role in this success, and what makes them unique is that they go beyond rooftop solar panels to incentivize other low-carbon technologies that relieve the pressure on the grid. Because they reduce consumption, air-source heat pumps and energy storage became a key part of Australia's renewable energy program.[82] The incentives offered to homeowners are significant. The system works like this: renewable energy certificates (RECs) are provided up front to small solar or wind generators, based on the anticipated amount of power that will be produced over a fifteen-year period. Most households grant these rights to an agent, who covers the cost of the system and in return receives a lower purchasing price.

In many Australian states, large energy retailers, or utilities, are required to purchase and "surrender" a certain volume of certificates each year to a government-run "Clean Energy Registry." Certificates can be bought and sold, so a market is created. The system appears to be highly robust. In 2023, for example, 28.5 million certificates were "surrendered" to the government's Clean Energy Regulator—after being purchased by retailers to meet statutory obligations.[83]

While Australia's feed-in tariffs went beyond solar, they initially focused on rooftop systems. At the peak of the subsidies, homeowners could earn as much as 60¢ (AU) per kWh, which was almost three times electricity rates at the time. Gradually, the rates dropped to around 8¢ per kWh today—still enough to pay for the initial investment in about five to eight years in most parts of Australia.[84] It is little wonder that, in 2023, one of three roofs in the country has a solar system; collectively, renewable systems generate more than 35 percent of the country's electricity.[85] Yet it is also ironic that, due to population growth, total greenhouse-gas emissions in Australia continue to rise.[86]

Just like everywhere else, the sun doesn't shine at night Down Under. As part of its efforts to achieve 100 percent net renewables by 2030, South Australia offers households up to AU$6,000 to install electricity storage, usually via lithium battery systems, through the state's "Home Battery Scheme."[87] South Australia's minister for energy and mining recently announced additional grants to help early adopters finance storage systems and become prosumers.[88]

As in Germany, Australian feed-in tariffs are ephemeral, usually granted for a limited period of time—what University of Queensland professor and sustainability advocate Jane O'Sullivan calls "a moving feast." As an early adopter in 2007 when photovoltaic systems were still exorbitantly expensive (AU$5–10 dollars a watt produced),[89] O'Sullivan received a AU$10,000 rebate for her initial investment in a PV system from the Queensland government. She also had access to a generous feed-in tariff rate of 45¢ (AU) per kilowatt hour of electricity. This generated a significant sum, but it did not take long before the program was discontinued.[90]

Even in the absence of tariffs, investing in solar panels and storage still makes good financial sense given the surfeit of sunshine in Australia and the high price of conventionally generated electricity. But, unfortunately, not everyone can finance solar panels or lives in a building which allows it. Community-based initiatives to make renewable systems more accessible for renters and disadvantaged populations are emerging, but they have not really changed the regressive nature of Australia's present energy system.[91]

Today, electricity accounts for roughly 25 percent of Australian fossil fuel emissions. With more than seventeen tons of CO_2 per person, Australia's per capita carbon footprint is still among the highest in the world.[92] Decarbonizing the economy will have to include parallel transitions for transport, heavy industry, and agriculture. But taxes and tariffs for renewables prove that change is possible. And that change can be very rapid.

Carbon Contracts for Differences

Given feed-in tariffs' broad success in promoting renewables, it is well to ask if similar strategies might jump-start other promising technologies. Carbon contracts for difference (CCfDs) are designed to do just that. As the name implies, governments use these contracts to encourage companies to adopt sustainable but more expensive technologies by covering the cost difference—*up front*. In a sense, the policy neutralizes any "late-mover advantage" for conservative companies that are only willing to deploy mature technologies.

Typically, a carbon contracts for difference policy proceeds according to three stages:

- First, a company or developer bids for a CCfD contract, providing detailed documentation of the anticipated expense associated with using a low-carbon technology.
- Second, the government responds by setting a "strike price" to remunerate emissions savings. This is paid up front, as soon as the environmentally preferable (and more expensive) technology is utilized.
- Finally, after adoption is complete, the actual costs are authenticated. If they are indeed higher than the strike price, the adopter keeps the strike price payment. If the costs turn out to be lower than the strike price, the "delta" is returned to the government.

Perhaps the greatest advantage of CCfDs is their ability to offer certainty to investors. The initial investment is lower, so the risk is lower, which prods companies to make a leap of faith and embrace a low-carbon technology without facing a competitive disadvantage. The policy constitutes good news for climate innovation and venture capitalists, as it can significantly shorten the lag before a new, environmentally superior alternative can compete with an existing, carbon-intensive technology.

CCfDs also have disadvantages. The primary downside of the policy is that it does not address uncertainties surrounding the actual operational costs associated with novel processes. This means that total costs for a CCfD will never be fully predictable, *ex ante*. Moreover, like feed-in tariffs, collectively these contracts are not cheap. To be effective, the government has to write out many sizable checks, and ultimately it is taxpayers who must guarantee a price that can catalyze a newer, more climate-friendly technology.[93]

Increasingly, governments are using this instrument in their efforts to meet carbon-reduction targets and encourage investment in cleantech. Early successes suggest that the strategy could be applied broadly across different areas of climate tech. As part of its "Sustainable Energy Transition Scheme," the Netherlands was among the first countries to establish a CCfD program. The Dutch program is designed as a "one-way" scheme: companies that receive subsidies do *not* have to pay any surplus revenues back to the government. The government contract, however, sets a cap for the total amount it will pay if the technology does not deliver, and if revenues fall below a specified level. CCfDs have been applied to support technologies involving renewable electricity, heat, and industrial technologies. Companies are typically selected based on the emission-intensity of their technology, with the contracts stipulating meticulous monitoring and reporting, via net-metering, once the subsidy is granted.[94]

Similarly, the German government is finalizing a CCfD program to support new ways to manufacture energy-intensive steel, glass, chemicals, and cement. Initial projections estimate that the program could reduce CO_2 emissions by 20 million tons a year. This comes to roughly a third of the targets set for industry by 2030 under the German Climate Action Law.[95] CCfD programs have also supported wind and biomass projects in the UK, carbon sequestration in Canada, and green hydrogen in the European Union.

These examples suggest that carbon contracts for difference programs hold great promise to accelerate climate innovation among large industrial sectors. This was indeed the trend in the first wave of contracts made with the private sector. There is no reason why CCfD programs cannot be employed in the same way that loans and feed-in tariffs were used to transform the economics of PV manufacturing: subsidizing individuals' early adoption of novel, low-carbon inventions.

Changing the Economic Calculus

The power of tax credits was first unleashed some fifty years ago to push consumers and corporations toward environmentally friendly electricity. At the start of the twenty-first century, Germany jump-started the widespread adoption of solar systems through feed-in tariffs, a policy that was soon successfully emulated in disparate countries around the world. More recently, carbon contracts for difference programs emerged as an effective instrument to incentivize a range of innovative technologies. They all tell the same story. Tax credits, rebates, feed-in tariffs, and subsidies work. When the economic calculus for adopting new low-carbon technologies is improved, the business sector responds. So does the general public. To catalyze innovation, smart money can be spent, temporarily, to subsidize new, high-priority technologies. Once demand increases, prices naturally fall, and so do carbon emissions, hastening a decarbonized future.

CHAPTER 6

Forcing Climate Technology

The earliest social and economic policy prescriptions did not come in the form of economic incentives or architectural nudges. They were command-and-control directives. When governments decided to initiate change, "Thou shalt not" was at the heart of most regulatory programs and interventions. Modern environmental history is no exception. To solve seemingly intractable environmental problems, bans and other binding design standards were employed to accelerate technological innovation—a process called "technology-forcing." These policies frequently took the form of standards that were unattainable without meaningful technological innovation.

Early environmental legislation in the United States relied on technology-forcing on several battlefronts. In retrospect, the US Clean Air Act Amendments from 1970 deserve credit for one of the most dramatic leaps forward in the history of air-quality management. Responding to an increasingly acute public health threat from air pollution, the American Congress launched a bold, technological transformation. Not surprisingly, there was pushback.

In a famous 1976 US Supreme case, Union Electric Company, a Missouri power station, sued the US Environmental Protection Agency (EPA) for approving a state air-quality plan that the utility claimed was impossible to meet. Union Electric argued that there was no way it could purchase low-sulfur coal at a reasonable price, that the cost of meeting the permit requirements would reach the astronomical sum of $500 million, and, critically, that "the equipment to remove sulfur dioxide from emissions had not yet been invented." But the Supreme Court ruled that these circumstances were irrelevant because that is *precisely* what Congress intended: rapid improvements in air pollution through technology-forcing.[1] Technical feasibility and costs were not supposed to be part of the equation.

By far the most famous technology-forcing success story from the 1970s and the advent of modern environmental regulation was the statutory timetable requiring new automobiles to install advanced emission controls capable of reducing air pollution from tailpipes by 90 percent within five years. When Congress enacted the rule, the internal combustion engine was a mature technology that had not been meaningfully modified in decades. Automobile manufacturers at the time claimed that the government's expectations were delusional and would lead to the shuttering of the American auto industry. But it then became clear that catalytic converters were perfectly capable of achieving such ambitious targets.

The invention itself is generally credited to Eugene Houdry, a French American engineer, who developed the first catalytic converter prototype in the 1950s. Yet it was not until the 1970 Clean Air Act Amendments and their emission-reduction demands that catalytic converters attracted much attention. The first-generation converters used pellets, or "beads," coated with metal catalysts to reduce exhaust emissions. The gas would pass over the beads, whose platinum or palladium reacted with the unburned hydrocarbons, converting them into water and carbon dioxide.

American automobile manufacturers realized that the government meant business. They subcontracted with Corning Inc. to develop a product that could convert car-exhaust pollutants into harmless gasses. A "catalytic regenerator," similar to converters used today, was eventually developed. Five years later, in 1975, catalytic converters became a standard feature in American automobiles, whose exhaust became much cleaner.[2]

Early catalytic converter models were considered a burden on vehicle performance and quickly gained a reputation as "horsepower killers." But the innovation created momentum, and the early models were soon improved. By the end of the 1970s, a newer generation of "three-way converters" was introduced to meet an even more challenging 1976 emissions deadline to reduce oxides of nitrogen. The next-generation device was more effective and better able to control a broader range of emissions, using rhodium to break down NO_x pollutants into relatively harmless nitrogen and oxygen emissions.

Development of three-way converters slowed, and auto manufacturers actually ended up missing the statutory deadline for adoption. Nevertheless, when the new 1978 Ford and General Motors models sported the technology, an industry tipping point was crossed.[3] Exhaust backpressure increased with three-way converters, but the overall effect on internal combustion engine performance was soon considered minimal. Over the subsequent decade, total emissions of conventional air pollutants from mobile sources dropped precipitously, despite a 34 percent increase in vehicle miles driven.[4] Today, as civilization witnesses the final stage of the internal combustion era, catalytic converters continue to reduce hydrocarbon and carbon monoxide emissions by about 90 percent, as well as diesel particulate matter emissions by 25–35 percent. This would not have happened without an audacious technology-forcing statute.

The effectiveness of technology-forcing as a tactic becomes clear when comparing American and European mobile-source pollution controls during this period. Today, the EU is an aggressive agent for environmental change. But this is a relatively new role on the world stage for the continent. During the twentieth century, European countries were far more hesitant about embracing environmental innovation. In the 1970s, the German auto industry resisted pressure by the government to adopt comparable emission standards that would have required catalytic converter installation. Small-car manufacturers in France and Italy (e.g., Renault, Fiat, Peugeot) also objected due to concerns that any additional costs would deprive them of critical market share.

In the absence of technology-forcing statutes, it was not until 1984 that European standards for HC, CO, and NO_x were reduced by 70 percent and catalytic converters began to appear in new cars sold on the continent. Even then, full implementation would be delayed by many years.[5] By contrast, Japanese car manufacturers, far more dependent on the US auto market, made the effort to meet the American standard.[6]

Some commentators at the time criticized policies that expedited catalytic converter technology in lieu of championing more fundamental innovations, such as "zero emission" alternatives to the internal combustion engine.[7] Given present knowledge about CO_2's insidious climate impacts, today such critiques appear prescient. Yet catalytic converters still offer a compelling case of "end of the pipe" technologies that were able to eliminate much of the world's most pernicious automotive discharges. In the United States alone, converters are credited with eliminating over 1.5 billion tons of air pollutants annually. Originally designed for cars, catalytic converters soon became available for use on buses, trucks, motorcycles, construction equipment, lawnmowers, and non-vehicle engines.[8]

There are several other examples throughout environmental history where technology-forcing produced a favorable outcome: California's 1990 requirement for 3 percent zero-emission cars ostensibly fell short of the state's deadline, but nevertheless is credited with triggering significant improvements in battery technology.[9] Bans on PCBs and DDT led to development of more environmentally friendly alternative chemicals.[10] Still, technology-forcing programs are not trouble-free. They face a range of political, economic, and practical problems. Even so, environmental historians can already point to a few climate tech–forcing success stories.

Three Tales of Climate Technology–Forcing

Mandatory Solar Water Heaters—A fifty-year retrospective: The first technology-forcing policy to catapult renewable energy forward took place in the 1970s. Geopolitical exigency was the proximate force behind the original legal requirement for passive solar water heating. In the 1950s, the recently established state of Israel found itself poor, diplomatically isolated, and chronically short of energy. In order to prioritize electricity for local industry, in 1952 the government prohibited heating water in homes between the hours of 10:00 a.m. and 4:00 p.m. Not surprisingly, many citizens were loath to forego hot showers during the day. The ban created a business opportunity to provide an off-grid solar alternative. Despite critics who bemoaned the aesthetic blight of the rooftop contraptions, Ner Yah became the world's first commercial solar energy company in 1953.[11] For the coming two decades, some three thousand passive solar water-heating units were sold each year in Israel.

There is nothing novel about the idea of solar water heating: in 1767, Swiss inventor Horace Bénédict de Saussure succeeded in trapping thermal radiation in a box with three layers of glass, producing temperatures that exceeded water's boiling point.[12] The initial Israeli passive solar systems were somewhat more sophisticated and did not reach such scalding temperatures. There have been incremental upgrades, but for seventy years now a cost-efficient model has been available for private consumers: an insulated tank holds roughly 160 liters of water that flows through a gravity-driven piping system. The water passes through a flat, two-square-meter glass panel, laid over a dark absorber plate with painted steel tubes. Water leaves the passive solar heater at a temperature of 50 degrees Celsius, a full 30 degrees higher than when it is pumped into the system.

Although corrosion eventually sets in, with no pump or moving parts the maintenance requirements are minimal. System life expectancy can easily exceed a decade, and energy-efficiency levels remain relatively high, typically reaching 50 percent. This means that a family installing a system on its roof can save 2,000 kilowatt-hours in electricity annually, for years. For many households, heating water comprises up to 40 percent of their total electricity bill, making the savings substantial.[13] Nonetheless, installing a system on an existing home requires capital up front, and putting a system on the roof of an apartment building is a hassle. As energy became more available during the late 1960s, and the prohibition on diurnal water heating was canceled, demand for solar water-heating systems began to dissipate.

This changed in the aftermath of the 1973 Yom Kippur War between Israel and its neighbors. Arab countries imposed an oil boycott, creating an acute challenge for Israel's energy sector. Israeli decision-makers sought ways to reduce electricity consumption. With a commercially available technology and an internal production chain already in place, in 1976 the government required that solar water heaters be installed on the roofs of all new buildings up to eight stories high—at that time, essentially all new construction in the country.[14]

The solar water heater requirement allowed the very modest local industrial sector to expand rapidly. Hundreds of resourceful welders and metal workshops offered up signature improvised models to meet the new demand. Eventually, five manufacturers managed to buy or crowd out competitors, cornering the market.

Rami Tarbulaki's father founded one of them, Nimrod Solar and Electrical Water Heaters, fifty-six years ago. Today, Tarbulaki continues to run the family-owned company. In his free time, Rami chairs Yisol, the trade organization that represents the solar water industry's interests. He explains that for many years Israeli solar water-heating companies exported units abroad, until prices in international markets became too low to justify the effort. With Israel's population growing at 2 percent annually, there is a large enough local market to keep 800 people engaged in production—and another 2,500 installing some 300,000 solar water heaters on Israeli roofs each year. As of late, high-rises are a growing percentage of housing stock in the country. But even in skyscrapers, the top seven floors are still required by law to heat their water with the sun.

Fifty years ago, with little regard to public opinion, a simple tweak of a building code forced a new technology on an entire country, creating a new industry and slashing the collective carbon footprint.[15] The efficiency of passive solar heating-systems slowly but steadily improved. Present estimates by industry representatives suggest that some 90 percent of Israeli homes have solar-heated boilers today, reducing electricity demand on the grid by 6 percent.[16] It would take Spain another thirty years to be the second country to adopt mandatory installation of rooftop solar heaters (2005), even as other governments opted for softer promotional programs such as tax exemptions (France, the United States), or subsidies (Germany, South Korea, the Netherlands, Sweden, etc.).[17] Had they taken the technology-forcing route, it probably would have saved their citizens a lot of money and significantly reduced greenhouse-gas emissions.

The Logic of LED Lighting: Today the world uses some 2,900 terawatt-hours (TWh) of electricity to provide people with artificial light. This constitutes some 16.5 percent of total consumption—a significant fraction, to be sure. Only a decade ago the percentage was far higher, with illumination consuming 19 percent of generated global electricity. This shows how relative electricity demand for lighting decreased, even as the population expanded and general economic prosperity grew, along with the increased illumination that affluence brings. In other words, the drop in electricity usage reflects a substantial global increase in the efficiency of lighting systems, with no meaningful decline in light quality.[18] This happened very quickly once technology-forcing policies kicked in.

Traditional incandescent bulbs work by heating a filament until it glows. Fluorescent lights use a gas discharge to produce light. Neither type is particularly efficient. By contrast, light-emitting diode (LED) lighting works by passing a current through a semiconductor material, which then emits light. The energy savings are considerable: the US Department of Energy estimates that LEDs use at least 75 percent less energy than incandescent and florescent bulbs. That's because so much of the energy reaching incandescent lights is lost as heat, while a far higher percentage of energy is converted directly into light in LEDs. Moreover, LEDs last twenty-five times longer—tens of thousands more hours—than incandescent lights that burn out after a few thousand hours.[19] By some estimates, LED Christmas tree strings can last a family's entire lifetime!

Yet another advantage of LED lights is their ability to be dimmed, with trailing edge dimmers, which are smoother and quieter than the leading-edge dimmers typically used by incandescent bulbs.

This means that users can adjust the amount of light—and energy—according to their actual needs. LEDs can also be connected without overwhelming the capacity of wall sockets. In short, LEDs are more energy-efficient, longer lasting, higher quality, and better at surviving extreme temperatures, shocks, and vibrations. All this translates into lower electricity bills along with sizable savings from reduced replacement rates.

The economic and environmental benefits of the technology did not go unnoticed. Many countries launched campaigns to convince their citizens to switch.[20] Some went further, forcing the public to embrace the new technology. In 2007, Australia became the first country to call for an outright ban on incandescent lights and compact fluorescents. Citing the energy savings and greenhouse-gas reductions that would result from the switch to LED lights, the phaseout became operational two years later.[21] The EU was soon to follow. In 2011, it imposed a ban on incandescents across the continent, parallel to a phase-in of LEDs.[22] A year later, Canada, Brazil, and Israel were on board, with the latter subsidizing LEDs to facilitate public acceptance. By then adoption had reached a global tipping point. Most countries in the world opted to join the transition. It is estimated that by 2025 some 50 percent of global lighting will use LED technology.

Notwithstanding the superior energy performance of the new technology, in 2021 overall energy consumption for lighting went up by 5 percent compared to 2020 (along with the associated emissions).[23] This was because of increased lighting in buildings, reflecting post-pandemic economic rejuvenation. Nonetheless, LED efficiency is a gift that keeps on giving.

What can be learned from this experience? As LED production began to scale economically, the technology became a no-brainer. The European Commission estimates that its decision to ban incandescent lights saves the equivalent of 15 million tons of CO_2 emissions, equaling the electricity consumption of 11 million European households. It also saves the European economy €10 billion. Lighting constitutes as much as 20 percent of an average household's ongoing energy expenses. EU economists estimate that Europe would have had to impose a tenfold tax on incandescents to attain the same level of behavior change.[24] Indeed, even this scenario might reflect considerable wishful thinking. Inertia is always powerful, and the elasticity of light bulb pricing is probably more modest than conventional theory assumes.[25] The public needed to have its arm twisted.

In fact, public support for the ban on incandescents proved to be surprisingly high. This can be attributed to a modicum of guilt that comes from using wasteful light bulbs, personal pride in being part of a righteous energy-conservation crusade, and the dramatic savings associated with the drop in LED purchasing price.[26]

Lighting is measured in units of light per kilolumen. As the statutory bans on conventional lighting kicked in between 2010 and 2019, the cost for LEDs dropped fifteen-fold. from $12.90 to 89¢ per kilolumen.[27] Projections for the future suggest that these trends will only increase. By 2030, the global LED lighting market size is anticipated to grow at a yearly rate of 10.5 percent, from $59 billion in annual sales in 2022 to more than $132 billion in sales by 2030.[28]

Per-unit purchasing price is sure to become even cheaper. People intuitively understand that there are times when forcing a technology on the public simply makes sense. LEDs are a case in point.

The Scourge of Single-Use Plastic: Plastics are a major contributor to atmospheric greenhouse-gas concentrations. In the United States alone, plastic production is responsible for the release of 232 metric tons of greenhouse gas—a level equivalent to 116.5 gigawatts of dirty discharges from the smokestacks of the world's coal-fired power plants.[29] In 2022, the US Department of Commerce found that plastic's carbon footprint was far greater than that of global aviation.[30] But this disturbing snapshot is not the whole story. The UN Climate Change Convention recently published projections that offer far more disconcerting reasons to worry: ". . . despite progress, plastic production is soaring, estimated to triple by 2050, when it will be responsible for 13 percent of the planet's carbon budget, *that is 56 gigatons of greenhouse*-gas emissions. . . ."[31]

A considerable portion of plastic's carbon footprint comes from single-use plastic bags. One single-use plastic bag is equivalent to 1.58 kg CO_2, or the emissions equivalent of driving five miles.[32] That's not much. But every year, the world economy manufactures 5 *trillion* plastic bags.[33] When the transportation equivalent is calculated, humanity's use of single-use plastic bags is tantamount to driving 40 trillion kilometers. In the United States, the world leader in annual distances driven, all told, drivers travel about 5 trillion miles each year in their cars.[34] That's only an eighth of the annual emissions from the worldwide manufacture of single-use plastic bags.

In short, plastic has fallen from grace as the material of the future: today, it is seen as a public-health and ecological nightmare. Only a tiny fraction, probably less than 5 percent, of plastic, is truly recovered.[35] That's almost entirely polyethylene terephthalate, or "PET," the kind of plastic used in beverage containers. PET can be melted down and spun into fibers for fiberfill or carpets, or even returned to its original use. Salvaging the vast majority of plastic, however, is far more vexing. For example, plastic bags, almost without exception, are simply not recyclable. Unfortunately, countless well-meaning citizens harbor the idea that the plastic bags they meticulously separate are recycled and become part of a circular economy, when they are in fact single-use, and part of the present crisis.[36]

In 2017, largely due to concerns about cleanliness in public spaces, Kenya became the first country to impose a nationwide prohibition on the use, manufacture, and import of plastic bags. But efforts to reduce single-use plastic and find low-carbon replacements are not just about litter control or protecting marine life.

Plastic bans also constitute a meaningful climate-change mitigation initiative.[37] The rule empowers Kenyan police to fine anyone carrying a plastic bag. Violators face fines as high as US$38,000 or four years imprisonment. Professor Judi Wakhung, the country's minister of environment, who championed the policy, assuaged public concerns by explaining that her main enforcement priority involved targeting manufacturers and suppliers, not ordinary citizens.[38]

Unsurprisingly, the phaseout was extremely unpopular among local industry representatives, especially as Kenya is a major exporter of plastic bags throughout Africa. After the law was enacted, Kenyan Association of Manufacturers spokesperson Samuel Matonda claimed that the ban would cost 60,000 jobs and force 176 manufacturers to close.[39] Nonetheless, the regulation withstood judicial review and the plastic ban was even expanded to include beverage containers inside protected natural areas.[40] Neighboring Ethiopia watched closely, and in 2019 also adopted a ban on single-use plastic.

These initiatives reflect mounting concern about plastic on the African continent. A decade subsequent to Kenya's plastic policy, Rwanda began to regulate single-use plastic bags in 2008. The Rwandan law allows for many exceptions, permitting single-use plastic bags for the packaging of meat, chicken, fish, and milled cassava leaves to allow for easy refrigeration. Although less comprehensive and stringent than the Kenyan rules, the Rwandan law contributed significantly to cleaning up the countryside. Fines may be lower (up to $300 per violation) but are high enough to be deterrents.[41]

Much like the LED regulatory story, the plastic-bag ban cascaded across continents. The EU soon followed the early adopters in Africa. In 2019, it passed the Single-Use Plastic Directive as part of its circular economy strategy for the continent. European countries like Italy, Belgium, and France soon set target dates for banning single-use plastic bags.[42] In the absence of a federal statutory prohibition, several North American cities such as Toronto, Montreal, and San Francisco also began to ban plastic bags locally. In Asia, China, India, and Taiwan have also set in motion restrictions on plastic bags.

The UK has gone even further: in 2023, it announced the imminent banning of single-use plastic plates, trays, bowls, cutlery, balloon sticks, and certain types of polystyrene cups and food containers. While there is a clear climate dividend, the aesthetics of litter control were front and center in the public outreach campaign. British environment minister Rebecca Pow characterizes plastic as "a scourge which blights our streets and beautiful countryside." She explains that before England banned plastic straws, stirrers, and cotton swab sticks, they composed almost 5.7 percent of marine litter. After the ban, a 2021 environmental analysis reported that "cotton buds" were no longer in the UK's top ten most common beach litter items.[43]

Here again, technology-forcing spurs climate-friendly alternatives. Biodegradable plastic bags, made from materials such as corn starch or cellulose, are becoming increasingly affordable. Under the right conditions, within a few months these plastic replacements can be broken down completely into water, carbon dioxide, and compost by microorganisms. It turns out that bags made from polylactic acid (PLA) derived from the sugars in sugar cane, cassava, or corn are essentially carbon-neutral—and edible![44] And paper bags are best for biowaste.

While scientists tout enormous environmental benefits from increased adoption of bioplastics made from plants such as corn (maize),[45] there are also downsides. Recent life-cycle analysis studies report that almost a quarter of agricultural land producing grains are utilized to produce biofuels and bioplastics. A rise in food prices can be expected as production increases to meet demand, which will disproportionately affect the well-being of low-income families worldwide.[46] The resulting conversion of rangelands to croplands can lead to greater than 50 percent depletion of soil carbon.[47] In addition, when traditional plastics made from fossil fuels were compared with bioplastics, not surprisingly the amounts of fertilizers and pesticides associated with the crop-based products was far higher as was the contribution to ozone depletion.[48]

At the same time, systematic comparisons invariably highlight the many eco-friendly characteristics of bioplastics. For example, production of PLA saves two-thirds of the energy needed to make traditional plastics with no net increase in carbon dioxide gas associated with its biodegradation.[49] Scientific research consistently validates the associated greenhouse-gas reductions: a team at Carnegie Mellon University found that a shift to corn-based biopolymers produced with *conventional* energy in the United States could reduce industry-wide GHG emissions by 25 percent, eliminating 16 million tons of CO_2 each year. If single-use plastic production were to rely on *renewable* energy, the drop would be far more precipitous, roughly 75 percent below present emissions.[50] In another life-cycle analysis, an Italian research consortium found that replacing single-use plastic items with compostable plastic and multi-use materials would cut CO_2-equivalent emissions by 73 percent to 90 percent.

The global biodegradable plastic market is starting to respond to the demand, as the public and industry slowly shift away from conventional single-use plastics. In 2021, global sales of biodegradable plastics were estimated to reach $4.4 billion. If present projections for 9.3 percent annual growth are realized, by 2027 the market for bioplastics will grow to $7.5 billion.[51] And yet this is not nearly fast enough. Technology-forcing can help fill the gap.

One company relying on forcing policies is ReSource Chemicals, the Berkeley, California–based company introduced in chapter 3. It is developing a bio-based platform for making 2,5-Furandicarboxylic acid (FDCA). FDCA as a basis for nonpetroleum-based plastic. Strategically, ReSource is counting on the multinational chemical giants to prefer environmentally friendly FDCA plastic. But given the highly conservative corporate culture in the petrochemical industry, this will not happen voluntarily. Experience suggests that progress will only be made when the plastic industry is forced to meet mandatory environmental constraints. Dr. Aanindeeta Banerjee, ReSource CEO, keeps a close eye on the evolving regulatory climate:

> Last year, California passed legislation that essentially says that 65 percent of single-use plastic has to be recycled and that all plastic sold needs to be compostable or biodegradable. Last month, England banned single-used plastic, starting in October 2023. Top consumer companies are starting to realize that they are going to have to reduce their use of virgin plastic dramatically by 2030. We fit right in there.[52]

Dr. Amy Frankhouser, the company's vice president for chemistry, explains that most of the public has no idea that only 2 percent of plastic today is actually recycled. "If we are serious about implementing a sustainable plastic policy, we are going to have to make plastics that are fully recyclable."[53] Single-use plastics, it seems, are not going to disappear anytime soon. But with newly available chemistries, spawned by uncompromising phaseouts—they could.

Technology-Forcing: Lessons from the Past

There is of course something viscerally satisfying about public policies where governments simply command industry to stop doing "bad things" and start pursuing better technological alternatives . . . and then, magically, it happens—overnight—solving a systemic environmental problem. Unfortunately, if technology-forcing were so easy, it would be far more common. This leads to the fundamental question: "When do technology-forcing policies make sense?" Luckily, valuable lessons have emerged from evaluations of past technology-forcing interventions that are relevant for climate policy.

There is no shortage of reasons why decision-makers avoid technology-forcing strategies. To begin with, they create adversarial dynamics between government and industry. Regulators essentially strong-arm the private sector, expecting firms to commit substantial resources to research and development. From industry's perspective, such policies can be interpreted as a declaration of war. In response to uninvited technological "fiat," corporations may well dig in their heels, devoting resources to pressuring regulators to delay, relax, or rescind the standards—rather than pursuing the prescribed technological panacea.[54]

Economists tend to be wary of technology-forcing solutions. If there is a compelling societal need to intervene in the market, they typically prefer incentives to motivate industry to find its own, optimal solution, balancing whether the benefits of an induced innovation warrant the compliance costs. Economic orthodoxy holds that a steep tax on a polluting process, input, or emissions can achieve the same goal with far greater efficiency. Mandating specific technologies may come at the expense of an alternative, superior technology or even delay its development.[55] And of course there is always the fear of leakage: with the world being increasingly "flat" and the economy increasingly global, a tough command-and-control requirement could push industries into other, more-lenient jurisdictions.

Taxes, however, are hardly a perfect instrument. One of the problems with Pigouvian taxes described earlier is that they tend to infuriate the public and can be an easy target for vested interests opposing a sustainable-technology transition. Politically, it is far easier for elected officials to force a clean technology on an unpopular industry than to impose a new tax burden on voters. Moreover, simply incentivizing better performance tends to favor existing technologies. Thoughtfully designed technology-forcing requirements can stimulate product and process innovations, even as they accelerate new technologies that are critical to improved environmental outcomes.

In 2005, Carnegie Mellon researchers David Gerard and Lester Lave took a deep dive into America's catalytic converter experience and suggested three conditions crucial to the success of a technology-forcing government initiative: First of all, the overseeing agency must have meaningful credibility as an enforcing agent so that the regulated industry will internalize that the government means business. Secondly, competitive pressures need to exist within the industry to drive firms' research and development. Startups or established companies need to have sufficient funds or profit margins to take on the associated research and development expenses. This creates a dynamic whereby companies see a benefit in finding innovative solutions and being early adapters. Finally, governments must be fairly certain that the innovation they are betting on is the "right horse." A compulsory, innovative technology must be effective, reliable, and devoid of unexpected environmental consequences and unreasonable costs, while being truly superior to what's already in place.[56]

There is plenty of game theory informing the politics surrounding technology-forcing programs. No longer "on the same team," industry and the government can quickly find themselves in a cat-and-mouse dynamic. The government naturally tries to convince profit-maximizing firms that it is worth their while to invest substantial resources in the research and development necessary to bring about a socially desirable innovation. But if this becomes too expensive, or if companies do not believe that improved environmental performance will increase demand for a new product, they are unlikely to up their R&D game.

In democratic countries, firms enjoy the option of lobbying legislators or other decision-makers. Experience suggests that companies will resist regulators' demands for innovation that cut into corporate profits. It is not uncommon for them to wield political influence to get government agencies to back off.[57] Since industries tend to have more information than government regulators do about their technologies and the costs of upgrading them or finding an innovative alternative, they enjoy a natural advantage in the battle over the "hearts and minds" (and votes) of the relevant politicians.

Thus, before government agencies take the leap into the stormy waters of technology-forcing policies, they should make sure to bring with them meaningful technological competence, which can counter the informational advantage that regulated industries typically enjoy. They also have to be confident that the public strongly supports the underlying environmental and social objectives behind their proposed policies, lest political leadership blink when faced with industry resistance. In the absence of a forceful regulatory position, industries may choose to "play chicken" and intentionally miss a deadline for meeting a new standard, daring the government to shut them down.

Recently, efforts to electrify home heating in Germany demonstrate the importance of making sure that political will is solid before trying to force technology. In response to the 2022 Russian invasion of Ukraine—and Germany's dependence on imported fuels that it revealed—the Green Party promoted a "boiler ban" designed to prohibit installation of most oil and gas heating systems in homes, starting in 2024.[58] Only new heaters running on 65 percent renewables or more were to be allowed, limiting consumers to heat pumps, biomass, or district heating. But politicians began to waffle when the media presented the bill as bankrupting the middle class. Eventually, the resolve of the members in the ruling government coalition faltered. The pushback was sufficient to convince the Bundestag to delay the ban until 2028, when it is to be tied to municipal heat plans. In an ironic twist, a wave of hysterical German consumers then rushed to purchase "fossil boilers" while they are still available, in order to avoid switching to more-expensive heat pumps in the future. [59]

Gerard and Lave also argue that in order for technology-forcing policies to spawn innovation, robust competitive pressures within an industry (to supplement regulatory pressure) are critical. If a firm becomes convinced that it can beat rivals to a newly mandated technology for which it will enjoy patent privileges, or that a new technology can become an industry standard, it might forego an evasive strategy and instead embrace the R&D challenge. That's another reason why having a global market matters. While it is unlikely that a government will shut *down* an important domestic industry that provides gainful employment to numerous citizens, a foreign company might be genuinely concerned about being shut *out* of an important overseas market. In short, if governments want to accelerate research and development, then government regulators need to convince industry that they mean business and that they will not relax standards or extend deadlines.

A "prisoner's dilemma" matrix can emerge around the technology-forcing dynamics. If all the relevant firms cooperate in flouting a government's technology-forcing demand on a given industry, then they might well avoid the additional R&D expense. If only a few firms eschew this sort of a boycott, then noncomplying companies may end up paying a very high price. And of course, the government should seek the ideal payoff for the public interest from the get-go, rallying all industry players to take on a new technological challenge.

Another insight involves the length of time allowed for developing alternative technologies. Given regulators' mission of reducing emissions and improving environmental performance, they have an intrinsic interest in a short compliance period. But given the snail's pace of basic research, shorter timelines tend to engender only modest and incremental improvements. For example, an "end of the pipe" solution might be designed to reduce emissions rather than more fundamentally redesigning processes. There is always the concern that if firms are forced to rush, an entire industry might get locked into an inferior technological alternative. Longer deadlines for phaseout increase the likelihood of significant technology breakthroughs. But they also bring downsides: industries will have an opportunity to exacerbate natural information asymmetries that exist between regulators and firms.[60] Ultimately, regulators need to balance the benefits of delayed emissions reduction against the environmental consequences of a longer compliance time.[61]

The best explanation for American success in forcing the hand of the auto industry technologically back in the day, is politics. Here again, there are lessons to be learned from the 1970s and the Clean Air Act experience: in this case, Congress decided that tying the US Environmental Protection Agency's hands was tactically advantageous. Legislators assumed that when faced with industry pressure, cautious bureaucrats will always opt to extend deadlines. So they decided not to give the US EPA the authority to postpone. Once executive branch bureaucrats and officials became statutorily disempowered, industry was faced with far less promising negotiation prospects. In addition, the law also threatened draconian penalties for missing the technology deadline: car manufacturers faced a $10,000 fine for every vehicle sold that did not meet EPA's certification test, this at a time when average car costs were only $5,000. Politicians gambled in order to increase the probability that the auto industry would move toward the R&D outcome—the optimal outcome in a prisoner's dilemma matrix.

It is interesting that this strategy worked for the initial 1975 requirements for catalytic converter adoption. But when the subsequent NO_x, three-way catalyst deadline came in 1976, the technology was simply not ready for implementation, given the necessary engine and computer technology upgrades. Congress quietly pivoted, agreeing to push the advanced emissions standard deadline back by several years.

The existence of a competitive global market that can participate in the innovation challenge makes for healthier regulatory dynamics. Firms should either fear that competitors will be one step ahead of them in the innovation race or see an opportunity to increase the costs for their rivals. International participation is also important, allowing firms around the world to share economic opportunities from meeting other countries' technological mandates. Case studies highlight the importance of predictability in negotiations. Technology-forcing rules should not be subject to frequent changes. Firms that take on financial risks to develop a technology are justifiably frustrated when governments move the goalposts. Without ensuring that these conditions exist, profit-maximizing industries may simply decide to grandstand and call the government's bluff.

Technology-forcing programs tend to scale existing, climate-friendly and cost-effective technologies. They may do little to catalyze new transformative climate tech innovation. This hardly constitutes a policy failure. In other words, technology-forcing programs may not force industries out of their comfort zones onto new technological horizons. Rather, their primary contribution may involve diffusing proven technologies that have not yet been become industry standards due to unfavorable economics, inconvenience, or innate human resistance to change.

Up and Coming Climate Tech–Forcing Initiatives

As governments consider their policy options for decarbonizing the economy, technology-forcing programs should be on the menu. Two emerging interventions show how governments can deploy climate tech:

- Requiring photovoltaic installations in newly constructed buildings and parking lots;
- Phasing out fossil-fuel-powered cars and trucks and replacing them with electric vehicles.

Reflecting bold decisions made at the highest political levels, both are starting to happen.

Germany, an early adopter of feed-in tariffs to get photovoltaics off the ground,[62] is starting to apply technology-forcing to deploy PV systems. Several German *Länder* (states) have begun to enact "solar obligations" in climate-protection laws. Cities like Hamburg and Berlin are also creating similar standards at the local level.

Requirements for solar installation on the roofs of new, nonresidential buildings and parking lots in Baden-Württemberg, a state in western Germany, came into force on January 1, 2022. The government enacted a rule mandating PV integration during roof renovations a year later. Such an edict suggests where Germany is heading.[63] Roughly 14,300 new residential buildings are built in Baden-Württemberg each year. Some 80 percent of the new housing stock have roofs where solar electricity generation is deemed to be feasible. Yet, before the new technology-forcing law kicked in, only 11 percent actually installed PV systems.[64]

In addition to the building requirement, Baden-Württemberg's "photovoltaic obligation" applies to open parking areas with at least 35 to 100 spaces in areas zoned as suitable for solar covering (e.g., not sited on slopes greater than ten degrees). The regulation requires that at least 60 percent of these car parks must be covered by photovoltaic modules. In order to meet the requirement, lot owners are also allowed to install alternative solar thermal systems or utilize external surfaces of adjacent buildings. Comparable statutes were subsequently passed in Bavaria, Rhineland-Palatinate , Schleswig-Holstein, North Rhine–Westphalia, and lower Saxony.[65]

As solar electricity becomes an increasingly inexpensive proposition, more countries are forcing their citizens to deploy photovoltaics. California recently imposed solar PV system requirements on newly constructed low-rise residential buildings.[66] All new houses in Tokyo built by large-scale builders will have to include solar panels after April 2025.[67] It appears as if the time for PV technology-forcing has arrived.

Another technology-forcing intervention that is starting to spread involves the *phasing in* of electric vehicles (EVs) in parallel to the *phasing out* of vehicles with internal combustion engines. With 30 percent of European emissions coming from the transportation sector, the European Parliament and Council agreed that electric vehicles were central to meeting the EU's 2030 pledge to cut greenhouse-gas emissions by 55 percent. In determining a 2035 deadline for a 100 percent cut in CO_2 emissions from new cars and vans (relative to 2021), the EU effectively declared that the final chapter for fossil-fuel-powered vehicles on the continent is at hand.[68]

Germany was already on board. After the Green Party returned to the ruling governing coalition as a full partner in 2021, it enthusiastically resumed its traditional "catalytic" role. The government wasted no time in announcing how seriously it took climate-change mitigation, setting 80 percent renewable electricity as its new 2030 target—a significant increase beyond the existing 47 percent level.[69] Germany's renewable-energy record has always been impressive. Transportation, however, is another story altogether.

Worldwide, 17 percent of greenhouse gases come from transportation; in Germany the percentage is 19 percent.[70] As opposed to other greenhouse-gas-contributing sectors where Germany is a "poster child," the country missed its recent targets for reducing transport-related emissions. Historically, the country's leaders have never felt comfortable setting any tough pollution targets that might stifle its powerful automotive industry. This appears to be changing. In March 2022, the German environment minister, Steffi Lemke of the Green Party, announced that as of 2035, only zero-emission vehicles will be sold in Germany.[71] Signaling the imminent end of gasoline-powered vehicles sends an important message to the public and to the private sector. Municipalities begin to establish charging infrastructure. The public begins to understand that there will soon be fewer gas stations in which they can fuel their cars. Assuming that the German electrical grid continues to decarbonize, making its fleet fully electric will dramatically reduce the country's carbon footprint.

In the United States, unsurprisingly, California took the lead in this area with a similar policy.[72] In 2022, the California Air Resources Board approved Executive Order N-79-20, which moves the deadline for transitioning to zero-emission vehicles up to 2035:

It shall be a goal of the State that 100 percent of in-state sales of new passenger cars and trucks will be zero-emission by 2035. It shall be a further goal of the State that 100 percent of medium- and heavy-duty vehicles in the State be zero-emission by 2045 for all operations where feasible and by 2035 for drayage trucks. It shall be further a goal of the State to transition to 100 percent zero-emission off-road vehicles and equipment by 2035 where feasible.[73]

Many countries have started their mobile source transition incrementally by targeting specific fleets. For instance, by 2024, all UK taxis must be electric.[74] By the end of 2025, all new buses bought in Israel must have zero emissions.[75] Much like the shift to LED lighting, it will not take long for many other countries to join the trend, albeit some of these may allow longer phase-in periods.[76]

Transformative versus Incremental Innovation

Harvard Business School professor Michael Porter wrote an influential *Scientific American* article in 1991 whose central premise came to be known as the "Porter hypothesis." Porter posited that strict environmental regulation encourages innovation in the form of cleaner and more-efficient technologies. These ultimately improve the commercial competitiveness of firms and lead to savings that more than compensate for the outlays for innovation and compliance.[77] By definition, technology-forcing policies should diffuse a requisite product or process. But the jury is still out as to whether technology-forcing rules will continue to produce fundamental transformations or simply mandate incremental improvements in technology efficiency.

By design, technology-forcing rules seek to accelerate technologies that can be feasibly supplied. While the catalytic converter story is inspirational, it may be anomalous. Technology-forcing laws are unlikely to spawn new energy sources, like hydrogen or nuclear fusion. For that to happen, legislators and regulators will have to pour funding into the research pipeline or set technology-forcing deadlines far into the future. There is, however, potential in other industries. Technology-forcing in the steel industry could include setting deadlines to switch to hydrogen-injection-based production.[78] The cement industry could be required to include reformulating cement chemistries or replacing clinker with supplementary cementitious materials.[79] As more data confirming the feasibility of pilot programs emerge, the risks associated with government interventions are lower.

David Gerard, the lead author of the pioneering article on technology-forcing, is a professor of economics at Lawrence University. Looking back, he confirms that legislatures can overreach in requiring unproven technologies. As an example, he points to the US Energy Policy Act of 2005, which mandated that domestic transportation fuel contain increasingly large volumes of biofuels, including cellulosic ethanol. In retrospect, the annual target of 36 billion gallons, set for 2022, was delusional. Due to a number of reasons (including technological setbacks and lack of financing), cellulosic biofuel production never reached a fraction of that level.[80]

Gerard believes that technology-forcing statutes can make their most obvious contributions by rapidly deploying technologies that are already mature. He gives two examples from the automotive safety world: shatterproof glass and airbags, which began as optional "add-ons" before they became compulsory. (Gerard observes that consumers tend to be more willing to pay for *safety* than *environmental* features in their cars.) He explains that corporations frequently have all sorts of clean technologies "sitting on the shelf" that they are loath to market because the economic return does not seem justified. These innovations are proprietary and so nobody knows about them. As a result, the public interest is not served. In such cases, only regulatory and competitive pressure can accelerate their introduction.[81] This situation is similar to "orphan drugs" in the United States (discussed in chapter 1), where relatively modest government support solved much of the problem.

An associated issue is the form that a technology-forcing standard should take. On the one hand, design standards can require adoption of a very particular technology: governments can set a design standard stipulating a low-carbon manufacturing process. Just as building codes can require solar water heaters on roofs or phase-ins of heat pumps for new buildings, design standards make sense when the objective involves scaling a proven technology, rather than catalyzing innovation per se. Alternatively, a regulatory approach can also set a performance standard. This implicitly gives firms greater flexibility to explore a range of solutions. Under the latter approach, governments remain fundamentally agnostic about how industries reach a goal. This minimizes the danger of "locking in" technologies that may prove to be inferior. By setting performance standards and then getting out of the way, governments may release greater creativity and more transformative outcomes. Depending on the technological challenge, there is room for both approaches.

The late biology professor and occasional politician Barry Commoner was one of the leading voices in the American environmental movement during the twentieth century. Commoner believed in unapologetically phasing out harmful technologies; scientists would ultimately find perfectly good alternatives. Forty years ago, he summarized his perspective in an article for the *New Yorker*:

> The real improvements have been achieved not by adding control devices or concealing pollutants (as by pumping hazardous chemicals waste into deep water-bearing strata) but simply by eliminating the pollutants. For example, the reason why there is so much less strontium 90 in milk and in children's bones is that we and the Russians have had the simple wisdom to stop the atmospheric testing of nuclear weapons which produce it. When the process that produces a pollutant is stopped—the banning of pesticides, the halt in atmospheric nuclear testing—there is considerable environmental improvement; if instead, an effort is made to control the pollutant by recapturing it or destroying it before it escapes into the environment, there is some improvement in environmental quality, but generally not much. In fact, such controls are ultimately self-defeating.[82]

Commoner's opinions were often disparaged as unrealistic, simplistic, or extremist. But there is a simple wisdom in rejecting a technology or a product with adverse environmental impacts and ingenuously saying: "*This is bad. As a society, we need to ban it and require something better.*" With climate anarchy threatening the future of humanity, the simple wisdom of forcing cost-effective, low-carbon technologies into the economy has an important place in the policy discourse.

CHAPTER 7

The Power of Public Procurement

In April 2023, prisoners at California's Folsom Prison lined up for breakfast only to discover that their usual bacon and sausage options had changed. Instead, they were served up a vegan breakfast burrito filled with a pea protein that had the flavor and texture of turkey. The California Department of Corrections and Rehabilitation framed its many new plant-based protein options as a health initiative, reducing the risk of heart disease, high blood pressure, and diabetes. The new foods were hailed as allergen free—and even kosher. "If I were incarcerated, I can't think of too many things that are more important to me than food, and maybe my mattress," explained Dr. Joseph Bick, director of health care services at the department.[1]

But as a matter of fact, the new menu had nothing do with health and everything to do with climate change. The transition to a more plant-based, locally grown diet was only one step that the Department of Corrections adopted in response to the state government's new procurement policies. The idea was to put the massive purchasing power of the state government to work in the race to decarbonize the California economy. State prisons would have to rein in their carbon footprints like everybody else.

A half hour away, in West Sacramento, a small but dedicated team of engineers was working to translate the state's new policies into specific rules.

Charleen Fain-Keslar had been running the engineering branch in the Procurement Division at California's Department of General Services for several years when the state legislature passed the Buy Clean California Act in 2017.[2] She was used to ensuring that the products and services purchased by government agencies were reliable, inexpensive, safe—and, ever since the Environmentally Preferable Purchasing Act was enacted in 2006, environmentally friendly.[3] But now the legislature expected her branch to regulate the emissions associated with construction materials and the embodied carbon in buildings. It was uncharted ground: an ambitious, climate-driven, procurement program. But she and her team dived right in. Soon they were flooded with calls from all over the world to learn about California's latest pioneering green initiative.

Under the Buy Clean California Act, the Department of General Services sets the greenhouse-gas emissions criteria, and then all state agencies have to implement them.[4] The Act requires that suppliers submit environmental product declarations to these departments for a range of materials, including rebar, structural steel, flat glass, and mineral-wool board insulation.[5] In practice, this means that prior to finalizing major contracts, contractors need to submit specific information about their products and processes. These declarations must demonstrate that the global-warming potential of their goods do not exceed the standards set by the experts in the Procurement Engineering Branch.

Setting these standards is not a trivial process. Some suppliers claimed that they would have to compromise trade secrets and make their products vulnerable to reverse engineering. But the government engineers held fast to the legislation, the standards got set, and the state agencies subject to the Act began to administer the law.[6] First the steel industry agreed to comply, then the rebar producers joined in, then the cement manufacturers got on board. The California government had a new, powerful tool to shrink its carbon footprint.

After the state legislature passed an amendment in 2022 directing the Department of General Services to address the embodied carbon emissions in construction materials, the regulators' attention focused on reducing the greenhouse-gas emissions associated with the building materials used for the construction of new government offices, buildings, hospitals, and schools. A year later a bill was enacted stipulating that a standard be set for the allowable global-warming potential associated with the cement, concrete, and asphalt used in state transportation projects.

Other agencies had to scramble to meet a bevy of new standards: alternative fuel requirements led to a shift to zero- and reduced-emission vehicles. Governmental departments moved quickly to diversify their fleets to include electric vehicles, renewable diesel, and a significant number of state-owned, hydrogen-powered vehicles. The computers in government offices now conform to some of the most demanding energy conservation standards in the world. In order to reduce recidivism, the Department of Corrections runs an entire Prison Industry Authority that provides on-the-job training so that inmates will have employable skills when released from prison. The Authority now has to meet climate-driven standards for manufacturing. California's prison population found itself, once again, on the front lines of America's most ambitious state-level climate-mitigation battle.

Public procurement makes a big impact because governments are big spenders. In Europe, governmental authorities use 19 percent of national GDP to pay for goods and services, while in the United States, around 15 percent of the federal government's $6-trillion annual budget goes to goods and services.[7] China is no different: in 2021 the government spent 15.9 percent of its $4-trillion budget on services and merchandise.[8] Beyond military outlays, most procurement supplies government-sponsored construction, health, and transport.[9]

Especially since it is financed by taxpayers, public procurement should start by guaranteeing fair and equal competition in the bidding for government contracts, ensuring that the best firm with the best price wins out. But it can go much further, serving the public interest by promoting social and environmental goals as well. Public procurement systems have come to embrace two different kinds of objectives, which can be defined as procurement and non-procurement. *Procurement* objectives normally involve, timeliness, cost, risk-reduction, competitiveness, and integrity. *Non-procurement* objectives include environmental protection goals, social priorities, and international relations.[10] Non-procurement objectives are frequently referred to as "green" public procurement or sustainable public procurement. The European Commission defines green public procurement as "goods, services, and works with reduced environmental impact through their life cycle."[11] Picking low-carbon products and services can contribute to decarbonization efforts in many ways: from prioritizing electric vehicles, reduced packaging, low-carbon cement, steel, and chemicals, to ensuring that wood products come from sustainably managed forests, or that food venders provide non-meat alternatives and reusable cups.

Procurement is also fundamental for businesses. But they are under no obligation to be transparent in their bidding process, or to abide by free-trade agreements.[12] Although there are multinational companies with budgets that are comparable to those of entire countries, most private-sector purchasing has more-modest dimensions. And while new ethics, sustainability, and governance

(ESG) standards are supposed to inform corporate behavior, for most companies, getting the best price from suppliers remains the priority. Size matters too: small firms do not have enough market power to incentivize suppliers to innovate.

For governments, however, procurement can be an effective *demand-side* policy instrument to support both the adoption and development of climate tech. Purchasing conventional products like office equipment, vehicles, street and indoor lighting, water services, heating, electricity—along with road and construction materials with a low-carbon footprint on a large scale can influence market dynamics. Governments can help stimulate production and drive down the price of these products.[13]

There are also cases where the government identifies a product or system that it needs, but that does not yet exist, whose design and production require novel, technological development . . . and then orders it anyway.[14] This is particularly important in areas like cement and steel, which contribute some 7 percent and 8 percent to global carbon dioxide emissions, respectively.[15] About 40–60 percent of all concrete purchases can be attributed to the public sector, and around 25 percent of revenues in the construction industry come from government contracts.[16] These statistics alone make it clear that governments can play a critical leadership role by pushing the private sector to break new ground on innovative products. At the same time, it is well to leave the design, creativity, and production to the private sector.

Government commitment to a new technology sends a signal: if it is affordable for the public sector, the private sector should also be considering it.[17] If implemented strategically, public procurement can contribute meaningfully to decarbonization. The following examples, however, show that this is easier said than done.

Tentative Steps toward Green Public Procurement

In 2008, a new framework began to affect the way local and national European governments purchase goods and services. Green public procurement criteria were established for twenty different kinds of products and services, governing everything from hazardous substances in furniture and pesticides in food to the energy-efficiency of buildings.[18] Then, in 2014, a European Union (EU) directive was approved that set minimum, legally binding environmental standards that need to be part of government's contracting process with companies.[19] That said, the directive only imposes its criteria on major purchases. For most projects, European governments are free to adopt their own environmental standards, but only if they choose to do so. And there's the rub.

The trouble is that the EU is a complicated place to pass mandatory rules. Lacking a clear consensus, green public procurement remains voluntary, with considerable discretion left in the hands of national and local officials. This explains its limited use in Europe.[20] Yet for those governments who do get with the program, buying green provides significant leverage. Life-cycle analysis that considers "cradle to grave" impacts can be applied to calculate the most "economically advantageous tender." Importantly, this may include the price of greenhouse-gas emissions and other climate change–mitigation costs.[21]

The timid spirit of the directive received a small shove forward in 2019, when the European Commission's Green New Deal was approved. The program includes a call on public authorities to lead by example and ensure "that their procurement is green" and that they mobilize industry for a clean and circular economy.[22]

While Europe is taking tentative steps forward collectively, several individual countries have already begun to demonstrate the enormous leverage that public procurement can have on emissions reduction. In the Netherlands, for instance, procurement by the government and public agencies accounted for some 20 percent of the country's entire GDP in 2020. In 2016, the government enacted a special law to encourage all agencies in the country—national and local—to join the procurement initiative. The statute is called the Public Procurement Act[23] (shocker!), and it presents a comprehensive system that covers a broad array of services and goods.

Even so, the Netherlands has only adopted a basic version of the EU public procurement directives, without additional environmental requirements at the national level. Moreover, it never appointed a national supervising agency with authority to monitor procurement programs. It still has a way to go before becoming a green procurement superstar.[24]

In 2005, a full decade earlier, the Dutch Public Procurement Expertise Centre or PIANOo (the Dutch acronym for "Professional and Innovative Tendering, Network for Government Contracting Authorities") was established, with a belief that "professional procurement can contribute to successful policy and offers value for taxpayers' money." After the passage of the legislation, PIANOo continues to operate under the Ministry of Economic Affairs and Climate Policy as a help desk for public authorities to assist them (and tenderers) in applying the substantive criteria in making procurement decisions.[25]

Although green public procurement remains voluntary in the Netherlands, some 170 public authorities have adopted a manifesto requiring sustainability criteria in procurement contracts. Most have yet to adopt an action plan.[26] Not only do most Dutch agencies prioritize environmental goods and services, but many also monitor the outcomes, calculating the greenhouse-gas reductions attained in avoided metric tons of greenhouse-gas emissions for eight product groups. When suppliers fall short of their climate-mitigation promises, poor performance is supposed to have consequences, such as a penalty 1.5 times higher than the original price discount. Yet until now, actual enforcement actions appear to be largely hypothetical.[27]

Unfortunately, cost remains paramount in most Dutch national procurement decisions, a clear indication of how far the national policy still has to go before truly delivering maximum climate dividends.[28] Local governments, however, are free to pursue more-ambitious procurement programs. And many do.

Rotterdam offers a fine example of how a city can be empowered by a national procurement policy. The municipal government decided that, by 2023, all the small cars owned or operated by the city would be zero-emission; all vans by 2025; and the entire fleet by 2030. The policy goes further: by 2025 all urban delivery of public goods and services in Rotterdam are to be zero-emission; internal moving services only have until 2024.[29] Such policies can be found in other European cities as well: Berlin attributes a 47 percent reduction in municipality-associated emissions to procurement of green commodities in fifteen product groups, while the Catalonia region in Spain adopted a 2030 target of 100 percent electric vehicles, while all public buildings will be fully powered by renewables.[30]

Slightly farther east, Lithuania has doubled down on green procurement. As of 2022, the country's online procurement dashboard reported that 53 percent of government purchases were based on green criteria.[31] But this is just halfway to where the country intends to go: legislation enacted in 2020 requires 100 percent green public procurements by the end of 2023, with only a few exceptions, like food procurement or defense needs.[32] As for oversight, the Lithuanian government prioritizes products with the highest share of emissions and the largest contracts. Although the country's national auditing agency expressed concerns about the high cost of green procurement, claiming it was "3.5 times more expensive than products without such characteristics,"[33] the Ministry of the Environment was undaunted. The ministry's comprehensive 2022 study of purchasing agencies suggests that green procurement is just getting started: 60.2 percent of the *value* of the public procurement in Lithuania was green—but in terms of actual quantity (in units), it was still only 12.6 percent.[34] Lithuania is designing monitoring programs for eighteen priority products, with particular attention paid to the energy and transport emissions of suppliers. When asked the reason for the country's ambitious procurement program, officials simply attributed it to the planetary need to move toward climate neutrality, as well as the local need to set an example for the private sector and society.[35]

Overcoming Public Procurement Flaws

Green public procurement is not without its critics. Some feel that these mandates are inherently unfair because small- and medium-sized companies, as well as firms owned by disadvantaged communities, are unable to compete.[36] This critique becomes even more compelling in programs that seek to stimulate innovation through the green procurement process. Others criticize inefficiencies that undermine the competitive process, which would otherwise yield lower prices for publicly funded projects.

Opponents also argue that green procurement quotas are overly prescriptive and sabotage competition, leading to higher costs for taxpayers. An OECD survey, for example, found that countries rarely consider the full opportunity costs and risks of procurement programs that seek to promote social or environmental objectives. Swedish researchers Sofia Lundberg and Per-Olov Markland assailed Europe's environmental aspirations in public procurement as leading to outcomes that are fundamentally inefficient and cost-ineffective. "Green Public Procurement is to be seen as a command-and-control environmental policy instrument, implementing direct controls. The result . . . gives less incentives to promote technological development compared to economic instruments such as emission taxes and marketable permits."[37]

The problem with this position is that it assumes that conventional tenderers without green criteria internalize the social cost of carbon. They don't. Most marketplaces today remain happily obtuse about any of their products' climate impacts. Granting commercial contracts to firms that impose costs on the planet actually sends a signal that "it pays to pollute"—and that the polluter *doesn't* need to pay.

A more legitimate concern is that many green procurement efforts have been feeble and their results underwhelming. Given the hundreds of billions of dollars available to influence the sustainability of goods and services offered throughout the world, public procurement's impact is largely disappointing. The Stockholm Environment Institute recently conducted research evaluating the effectiveness of the EU's green public procurements in two key sectors: construction and road transport. As of 2023, these sectors represent 12 percent of the greenhouse gases associated with government procurements in Europe. The institute conducted case studies for eight diverse European countries: Sweden, the Netherlands, France, Germany, Estonia, Poland, Spain, and Italy.

The study found that when requirements to prioritize low-carbon transport and construction suppliers are nonbinding, the overall impact is anemic. While there is no shortage of rhetoric, promulgation of sustainability criteria, and pilot successes, the report criticizes the programs' lackluster implementation. The authors conclude that green procurement fails to operate in a "focused and consistent manner, and with the scale required to harness the decarbonization potential of this tool."[38]

There are certainly ways to change this. One American-based NGO, Open Contracting Partnership, was established precisely to improve procurement performance and is now active in fifty countries.[39] Open Contracting Partnership documents the benefits of transparent procurement policies and raises awareness about implementing them. The organization identifies several obstacles to attaining meaningful environmental impact from procurement policies:

- *An insufficiently enabling environment.* Not enough public contracting authorities operate according to binding statutory frameworks that require them to actively prioritize green criteria.
- *The prevailing "tyranny" of lowest-price bidding.* Suppliers are incentivized to submit bids that neither internalize the true costs of the offer nor reflect the contract's actual climatic, environmental, or social impacts.
- *Risk aversion.* Public procurement rules are cautious and compliance-based because public officials quickly learn to avoid taking chances.
- *Inadequate capacity.* Most overseeing personnel lack the necessary expertise and training to understand the vagaries of climate change–mitigation and engineering specifications.
- *Deficient data.* Public procurement processes are still largely paper-based, which make data inconsistent and unreliable for evaluation. Fewer than half of the countries surveyed report on their procurement spending levels and write public contracts in a "machine-readable" format.[40]

The organization suggests an eight-point plan to address these obstacles, including expanding the rules favoring "planet-friendly" procurement so that they are comprehensive and a mandatory feature of all contracting processes at public institutions. Climate and environmental criteria should be built in as part of the core technical specifications and award criteria, forcing bidders to include environmental costs in their offers. Open Contracting Partnership also touts the importance of training via capacity-building workshops, consultations, and roundtables for both procurement practitioners and suppliers. Governments should establish dedicated units to coach and support public contracting authorities as they try to make informed decisions in areas that quickly become very technical in nature. In many places, reaching out to suppliers, through premarket consultations, roundtables, and competitive dialogues, has produced better outcomes. The learning process is quickly reflected in more-effective procurement criteria, getting suppliers focused on the key environmental improvements that governments would like to promote.

The city of Portland, Oregon, is an excellent example of a municipal procurement program that reflects Open Contracting Partnership recommendations. In 2019, it established a low-carbon concrete initiative requiring any city construction project to submit its cement to the city's testing lab. The rule ensured that publicly purchased concrete's global-warming potential is below the city's maximum allowable level within its product-specific strength class.[41] It turns out that cement contributes 25 percent to the construction sector's total emissions. Most of the cement utilized by the city—around 23 percent—can be "scientifically targeted" for abatement, and subject to technical specifications as a condition to receiving a contract.[42]

Like any environmental policy, public procurement programs should be driven by clear, quantifiable outcomes, or what have been called "science-based targets." For example, a city can decide that it wishes to offer 90 percent vegetarian food in public spaces, or to continuously reduce the embedded carbon in its new buildings by 10 percent every year. Alternatively, it can adopt a target of zero-emission deliveries. For instance, Helsinki, Finland, has begun setting such clear goals: it has mandated that 30 percent of procured buses be electric by 2025. In addition, a 74 percent reduction is to be achieved in emissions associated with heating in the Helsinki residential district.[43] Alternatively, an agency can simply adopt more general, sweeping objectives (e.g., reducing the carbon footprint of procurement activities by 60 percent). Thus the city of Rotterdam set a goal for 2020 that 25 percent of all public products and services be circular, producing zero waste during their life cycle, with that rate reaching 100 percent within the subsequent decade.[44] In either case, agencies should select key performance indicators so that evaluation can take place in real time. Programs need to be data-driven so that it is possible to monitor progress. Finally, Open Contracting Partnership suggests patience: "Go sector by sector. You don't have to try to do everything at once."[45]

Public Procurement and Innovation

It can be argued that public procurement is not the most natural way to cata-lyze innovation, since government bureaucrats tend to be a conservative cohort. They are well aware of being monitored by their superiors, supervisory agencies, the press, and the public. This makes them wary of spending public funds on an unproven technology. Benefits from innovative goods and services also tend to take a while to manifest, typically beyond the tender deadline—not to mention an upcoming election cycle. Procurement officials are already overwhelmed by the dizzying details and technical specifications involved in green procurements. Adding innovation to the criteria requires even greater expertise.

Often suppliers sense that contracting authorities are hesitant to make a leap of faith and opt for an innovative product. As they are liable for any risks associ-ated with trying something new, tenderers are acutely aware of a public sector bias for goods and services that rely on a familiar technology. If they can submit a competitive bid with a conventional product, many firms are disinclined to invest in innovative solutions.

Some experts, however, are more sanguine about procurement programs' potential to catalyze better technologies. Mark Dutz and Dirk Pilat from the World Bank and the OECD, respectively, cite three reasons why they believe that public procurement can pull innovation forward. By bundling together the demand of myriad agencies, commercial risk for innovation is reduced. To begin with, the government, as a lead user, can enhance the reputations of innovative businesses that win procurement awards, and the contracting process can cre-ate a dialogue between governments and relevant companies. Secondly, public buyers can provide advance notice about innovations that they seek to promote via procurements. They can also help expand the pool of potential tenderers or increase the quality of the goods and services. Finally, after identifying a key technological need, governments can use the procurement process to create a powerful market with enough muscle to motivate firms to conduct meaningful research and development.[46]

These are just some of the reasons that governments have begun to encour-age innovation in tenders. In 2014, the EU formally identified procurement's potential to catalyze innovation: "Public authorities should make the best stra-tegic use of public procurement to spur innovation . . . [which] plays a key role in improving the efficiency and quality of public services while addressing major societal challenges."[47] In 2023, it inaugurated a Strategic Procurement Dialogues project, with the goal of making public procurement a lever for innovation and sustainability.[48] Some countries are paying attention. German law specifically recognizes the importance of enabling public buyers to ask for innovative solutions in public tenders. France adopted rules that offer a price premium for innovative solutions in tender calculations. South Korea grants immunity to public buyers for losses that are not covered by insurance. And the UK established procurement procedures to reduce risks for procurement in the area of health management.[49] In one project, for instance, low-carbon illumination in a hospital was promoted through a procurement call more than two years prior to when such lighting was actually required (without specifying the desired solution), giving ample time to stimulate the entire supply chain.[50]

Such efforts seem to be working. A research team from the University of Manchester conducted a survey of 800 suppliers to the public sector in the UK. The study asked them whether they believe innovation affected the outcome of the tenders in which they participated. No fewer than 67 percent of relevant respondents indicated that bidding on or delivering contracts to the public sector made them more innovative. A full 25 percent reported that their innovations were almost entirely driven by the public procurement process. Furthermore, an impressive 56 percent of respondents believed that during the previous three years, the reason that they had actually won a contract with a public authority was due to their ability to innovate. Of those suppliers who carry out research and development programs, 51 percent reported increasing R&D expenses as a result of bidding for public sector contracts or delivering public goods and services. Finally, innovation is perceived as a winning strategy: more than 75 percent of suppliers reported that innovation helped them win additional contracts and increase their sales in the domestic private sector, with 29 percent reporting that innovation enabled them to win contracts overseas.[51]

Subsequent to the passage of the 2015 Paris Climate Agreement, the OECD conducted a review of member countries' experience in the area of public procurement and innovation. It recommended three ways that countries can harness the procurement process to boost climate innovation. The first is preliminary: identifying unsatisfied environmental needs and designing tenders so that a market can emerge to respond to them. If a country wants to induce innovation, size matters. That's why the second stage involves aggregating demand from various agencies across a country to create economies of scale sufficient to produce a significant market signal for a new product or service. Finally, the OECD's three-part strategy calls for providing a "reputational boost" to the selected product or process, with governments using their stature and resources to publicize an ambitious effort.

Local governments can also facilitate innovation with partners in their region and beyond. Innovation can address a local government's needs in collaboration with the private sector,[52] and by being the first to buy, local government can act as a lead user. Governments have the capacity to risk trying technologies that may not yet be fully optimized. And by seeking solutions to real-world problems, they can provide valuable opportunities for companies.

Creativity is something that governments can encourage directly. It happens all the time. For instance, innovation contests (also known as "hackathons") reward imaginative solutions to any number of challenges—including climate mitigation. Here again, central or local government agencies can create demand for innovative solutions by committing up front to be the first customer to buy and apply a novel product or process that wins the competition. Climate innovation technologies competed for recognition in two such programs in Tampere, Finland. These programs are credited with creating an interactive process and mutual learning between entrepreneurs and the government. The conclusion: by bringing together disparate actors in a focused competition, hackathons can produce climate solutions that are better suited to actual local conditions.

The 2021 Cambridge University review that appeared in *Nature Climate Change*, evaluating different climate policies, was extremely positive about government procurement's potential to fuel innovation. The authors conclude

> Government procurement can also promote innovation and competitiveness in small firms. Importantly, it can be used by local and regional jurisdictions to target localized problems for firms, such as facilitating access to new markets. In some cases, the use of public procurement in local and regional jurisdictions resulted in positive competitiveness and innovation outcomes through the creation of market opportunities, particularly for small and medium-sized firms in economically stressed areas. Given the challenges smaller companies face in getting novel sustainable products into the market, government procurement programmes should include flexible design features.[53]

The First Movers Coalition

As COP27 approached in 2022, US special climate envoy John Kerry proposed a new initiative to spur innovative technologies, saying, "We have to accelerate the clean energy transition and, my friends, it takes money to do that."[54] Thus the First Movers Coalition (FMC) was born.

Distilled to its essence, the initiative is a public-private partnership under the umbrella of the World Economic Forum. It brings together nine countries, including India and Sweden and, as of 2023, sixty-six of the world's leading corporations. The governments provide the publicity and moral support to "celebrate the climate leaders" from the private sector—while the companies pledge to produce new climate-friendly products. Corporations do not always keep their promises, but the organizers in the White House assume that the reputational damage for reneging will ensure that companies are as good as their word. With offices based in Geneva, Switzerland, the coalition is a formidable economic force: the collective market value of its members reaches $8 trillion.

Naturally, the initiative has not been free of criticism from academics, NGOs, and even labor unions, which have expressed dissatisfaction about inadequate ambition, lack of concern for developing countries, and insufficient stakeholder participation. And there are those who claim that it addresses the symptoms rather than the causes of the climate crisis. The coalition may not be perfect and remains a work in progress. But for technology start-ups and investors, it nevertheless is hard to overstate the importance of the world's major corporations, from Amazon and Google to Volvo and General Motors, openly prioritizing low-carbon technologies even when they are a little more expensive.

Orchestrating the First Movers Coalition is Dr. Varun Sivaram, a solar maven and sustainability prodigy. Sivaram completed his doctorate in physics at Oxford at age twenty-five and then went on to advise government bodies like Los Angeles and New York State about renewable energy policy. He was not yet thirty when his book *Taming the Sun* became the definitive word on scaling solar power.[55] It was already his third book. When Joe Biden decided to restore American leadership to the area of global climate mitigation, he drafted former Secretary of State John Kerry to coordinate international policy. Kerry put together a climate policy dream team, and Sivaram answered the call.

According to Sivaram, the fundamental assumption of the FMC's approach is to engage the "hard to abate" emissions sectors to make sure that "if you build a lower-carbon alternative—they will indeed come."[56] Sivaram was impressed by the success of advance market commitments in bringing healthcare technologies like Covid-19 vaccinations to the market in record time. He assumed that the same approach could push the climate tech needle. Under the FMC program, the US federal government and the private sector commit to purchase a specified product by a clear deadline. If the company meets the performance threshold by the deadline, the FMC investors will purchase the product. Sivaram offers a romantic vision of a "beachhead market" where investors and suppliers "make the leap together across the 'valley of death' and touch the pot of gold at the rainbow."

One particularly noteworthy gambit involves steel, which constitutes 8 percent of global greenhouse-gas emissions.[57] Ford Motor Company was among the first and boldest of the Coalition members, promising that 10 percent of its steel supply chain would be low-carbon by 2030. Automakers Volvo and Scania, as well as fifteen other major steel consumers, were quick to follow. Given production constraints, if the percentage commitment were higher, it is unlikely that manufacturers would be able to deliver the quantities demanded.[58] The First Movers Coalition strategy allows future steel supply to get built out while the new subsidies provided from the Inflation Reduction Act start lowering the price of low-carbon steel, making its production increasingly competitive. Eventually, purchasers should be willing and even delighted to purchase the low-carbon steel.

Nevertheless, many corporations prefer to stick with a familiar product. Pushing commitments into the future can help overcome such psychological barriers.[59] When recruiting companies, the American government appeals to companies' competitive advantage: once tough carbon standards are mandatory and the pressure to meet net-zero targets becomes significant, earlier adopters will be in a much better place in terms of their supply chains.

The program makes very serious demands upon its participants. It is not looking for "incremental innovation." Rather, its standards force technology forward; Sivaram avows that every technology needs to "bring something new to the market." For example, when the FMC was setting a standard for cement production, it was not enough to simply swap out dirty energy for clean energy in the cement production process. That's "something we already know how to do," says Sivaram.[60] Rather, companies will have to rely on carbon capture, utilization, and storage.

The strategic goal in focusing on 2030 products is bridging the technological gap as industry begins the final leg of its race to net-zero 2050 emissions. If firms are able to adopt a zero-emissions technology for 5 percent of production by 2030, it is fair to assume that within twenty years they will scale to the entire industry. Sivaram explains that if Ford gets to 10 percent "near-zero-carbon steel by 2030, it will not only have moved closer toward its 2050, net-zero objective. It will also make clean steel cheaper for every consumer on earth." The subsidies provided by America's newly enacted policies serve to de-risk and essentially eliminate the "green premium" that novel technology usually forces buyers to assume if they want to buy a novel, low-carbon technology.

Reporting and Rippling

Mariska Verseveld is a senior attorney working at PIANOo, the Dutch Public Procurement Expertise Centre. Although she came to her present position as a compliance officer, she recognizes that procurement policies often require a different approach. The procurement game needs "carrots" but also "sticks." When dealing with voluntarily green public procurement, the challenge is convincing government officials and companies to move toward the sustainable products and services they *both* would like to see. Many cities in the Netherlands are already active in pushing the market, calling for innovative products with better environmental outcomes. But they are still on the hook for ensuring that public resources aren't wasted on expensive products. This sends a confusing message. Suppliers still lack the confidence that if they develop the green products the government says it wants, they will be able to sell them.[61]

Verseveld has seen that when a sustainable product meets the price and functional standards of a major agency, it can create a ripple effect. For instance, it turns out that there are relatively few players in the Dutch furniture production ecosystem. This made it easier for contracting authorities to mobilize manufacturers and move them toward a cleaner, more circular menu of products for all consumers. Even though the dynamic is not something that has been measured yet, she knows it to be very real. Were the government to require all public bodies to apply green procurement standards, including monitoring and enforcement, she believes that low-carbon alternatives would ripple across agencies, municipalities, and ministries.[62] In this digital age, getting the word out should not be all that hard.

This intuition is supported by the South Korean experience. In 2005, the Republic of Korea passed the Act on Promotion of Purchase of Green Products. The law requires that all government agencies submit an annual green public procurement implementation plan. Targets are voluntary, but having to report each year creates expectations and momentum. Reporting is easy, so everyone complies. By 2017, all of the 910 organizations required to submit their procurement results, did so. The numbers are impressive: in a decade, total public outlays for green products grew by an order of magnitude, from $759 million in 2006 to over $3 billion in 2017. Roughly half of the total expenditure in product categories was classified as environmentally friendly.[63]

Governments can also expect their suppliers to report on their climate performance. In November 2022, the Biden administration proposed that all major federal contractors disclose their greenhouse-gas emissions and develop carbon-reduction targets.[64] The rule would cover an estimated 85 percent of the emissions associated with the supply chain of the federal government. Suppliers receiving as much as $7.5 million in government contracts would have to report their direct emissions along with those associated with the electricity they use. The *Washington Post* writes that such a seemingly small revision of the Federal Acquisition Regulation and other procurement guidelines "could ripple across the US supply chain."[65]

In order to do its job, government has to spend money—a lot of money. Procurement policies have passed the audition as voluntary programs; now it is time to make them mandatory for local and central governments. There is no reason why governments should not spend the public's money on climate-friendly products for the public good, saving money in the medium term and the planet in the long term.

CHAPTER 8

Nudging Down Carbon

In 1986, Schönau, a small German village near the Black Forest, was traumatized when a nuclear reactor core 2,000 kilometers away in Chernobyl, Ukraine, exploded out of control, spraying radiation throughout Europe. Families all over the continent were forced to stay indoors for days. Ursula Sladek, a homemaker, along with her five school-age children, was among those locked in by the contamination. The children were unable to play outdoors for two weeks. Even now, well into the twenty-first century, mushrooms in the nearby forest are still considered unsafe to eat.[1] When the radioactive isotopes finally abated, Sladek resolved to do something to change the situation.

She decided to "act locally," promoting alternative energy sources to the nuclear plant powering her community. It took no time for Sladek to face fierce resistance from the managers of the town's utility. So she simply decided to buy them out. Sladek launched a highly successfully fundraising campaign, and the money donated was enough to purchase the local grid. By 1997 she had established Schönau Power Supply, a new electricity cooperative, and assumed the role of president.[2] It was then that she implemented what was probably the first documented program that demonstrated the potential of "nudges" and choice architecture (influencing choice by organizing the context in which people make decisions) in order to decarbonize electricity.

Among the company's first decisions was to change the default option appearing in the bill that customers received from "gray," conventionally generated electricity, to a green alternative, representing renewable electricity. In those days, renewables were far more expensive than other sources of energy, so consumers were given the choice of "opting out" and going back to using the polluting but less expensive gray alternative. Almost without exception, 99 percent of the people of Schönau stayed with the default green-electricity alternative. Less than 1 percent "opted out."[3] Although Sladek had never heard of nudges, she had stumbled on an important public policy instrument.

It is not clear where the English word *nudge* comes from; some say its origins are in Scandinavia—as in the Icelandic word *nugga*, or "jostle." Others point to the Yiddish *nudnik*—referring to a persistent nagger. In modern policy jargon, *nudges* are a type of policy intervention that gently pushes people in a particular direction in the hope of getting them to make a desirable decision without depriving them of the freedom to choose. Their effectiveness derives from the observation that human preferences are malleable, and frequently they are a construct of our choice environment. Dozens of empirical studies have confirmed that nudges are statistically predictable in eliciting a particular behavior.

Policy makers need to devote more attention to addressing the so-called green gap—the pervasive disparity between people's genuine *concern* about the climate crisis and their actual *behavior*.[4] A range of emission sources that are directly linked to individual behavior give rise to massive greenhouse-gas emissions. These include diets that rely on animal-based products, addiction to "fast fashion," profligate electricity usage, dependence on gasoline-powered private vehicles, and large family sizes. Collectively, these emissions are 41 percent of the total US greenhouse gases released into the earth's atmosphere.[5] Getting to net zero is not just about influencing industrial activity and technological innovation. Personal decisions matter, too. For instance, what people eat matters a lot: about one-third of anthropogenic carbon emissions is attributed to the global food system.[6] As the decarbonization of electricity progresses and its emissions decrease, the percentage of greenhouse gases coming from the way people live their lives will increase.

Policies that seek to limit individual choice and autonomy in basic areas of private life are usually doomed to unpopularity—if not outright failure. That's because they often infringe on accepted societal lifestyles, entrenched normative behaviors, and even human rights. Most climate-mitigation policies, for example, focus on reducing the carbon intensity of electricity *supplied* by large-scale power plants, rather than the amount of energy *demanded* by their customers.[7]

But mounting consumer demand is invariably a big part of the problem. Here, the public needs to be prodded toward better consumption patterns. By utilizing choice architecture—a "soft path"—nudges are analogous to a gentle push that respects individual freedom and still allows society to pursue its collective objectives.

In general, its advocates do not purport that nudges offer a panacea for expediting full decarbonization. Nonetheless, as the case of Schönau's electricity supply shows, nudges can be a valuable supplement to conventional incentives or directives. This is especially true in the sensitive space where governments need to target individual behavior.

Released by Yale University Press in 2008 and reissued in 2021, *Nudge: Improving Decisions about Health, Wealth, and Happiness* was a publishing sensation, selling over 2 million copies. Highlighting research by two University of Chicago social scientists, the book offered dozens of examples of effective nudges that pushed human behavior toward socially desirable alternatives, solidifying the status of behavioral economics as an important area of research for public policy. Author Richard Thaler went on to win the Nobel Prize in Economic Sciences. His coauthor, Cass Sunstein, took a senior government post as head of the Office of Information and Regulatory Affairs in the Obama administration and later joined the Harvard Law School faculty. Overnight it seemed, nudges had become the "flavor of the month" in public policy: an innovative, noncoercive instrument in the policy toolbox, one that is informed by a "libertarian paternalistic" ideology that appealed to progressive and conservative politicians alike.

While they did publicize the concept, Thaler and Sunstein did not "invent" nudges, nor even the underlying psychological theory behind them. Historically, policies designed to change human behavior reflected the traditional assumption among economists that humans are rational actors. During the final quarter of the twentieth century, however, empirical evidence increasingly confirmed that people are anything but rational.[8] Often they are influenced by a range of biases and cognitive constraints.[9] Rather than trying to correct what appear to be natural human characteristics, policy makers began to recognize—and exploit—a range of heuristics, or rules of thumb, that typically guide individual reasoning and decision-making.

Changing the context, or the "choice architecture," in which people's decisions are made can encourage socially desirable behaviors. For instance, most people have an innate desire to keep their promises. While not legally enforceable, policies that elicit pledges and commitments can lead to outcomes as significant as those forced by fiercely policed command-and-control prescriptions.

Concern about reputational damage for many people is far more powerful than the unlikely prospect of facing an official enforcement action. For instance, when people were asked to make public commitments to consume less meat, it led to 15 percent greater reductions in meat consumption than individuals who were merely given information about its negative ramifications.[10] Similar impacts were found when people made pledges to utilize public transportation or to hang up towels for reuse in hotels to save water.[11] Policies can be designed to expose the public to social norms, expectations, and peer comparisons, which can consciously, and often unconsciously, change individual choices. Indeed, in another towel study, when guests were told that other guests were utilizing their towels more than once, it significantly increased their reuse.[12]

The rationale of nudges goes back to the findings of Princeton psychologist Daniel Kahneman about how people think. These insights were sufficiently profound to make Kahneman the first psychologist to win a Nobel Prize as a founder of the field of behavioral economics. Kahneman identifies two cognitive systems. The first operates automatically, whereby people have little ability to assert mental control. Examples of System 1 cognitive capabilities include recognizing familiar people, detecting disappointment in someone's voice, receiving messages from billboards, or evincing alarm at a charging lion. By contrast, System 2 is reflective, requiring exertion of mental effort to analyze situations and make sense of them. People use System 2 capabilities when making numeric calculations, monitoring their own or other people's behavior in a group, parking in a narrow space, and choosing to get married. (One can only hope!). System 2 thinking takes System 1 intuitions and turns them into meaningful beliefs and conclusions, upgrading impulses into well-considered behaviors.[13]

Unlike incentive-based or rule-based programs that promote deliberative and conscious thinking about behavior, nudges are designed to take advantage of superficial System 1 thinking. By identifying and exploiting the cognitive biases that influence people's decision-making, they can gently push the public toward a particular behavior without forcing it. One common way to nudge is through "choice architecture"—in which the environment informing a decision is carefully designed to produce a certain outcome. A classic example of this kind of nudge is the presentation of healthy food options at eye level at a cafeteria. This makes it easier for customers to select healthy (or low-carbon) options, requiring greater mental and physical effort to select a more remote and less nutritious item.

In public policy, nudges often create a "default" whereby choice architecture makes it easier—or even automatic—to select a certain behavioral outcome.

Perhaps the best example of how effective defaults can be is in the area of organ donations. There is a far greater demand from people who need transplants than there are available organs. In the United States alone, at any given time more than 100,000 Americans are waiting for organ transplants; more than 6,000 die each year while they are waiting.[14] That is because 80 percent of transplant organs come from people who are no longer alive. The tragedy is that, statistically, a mere 1 percent of donors' wishes to have their organs harvested after death are actually honored. Donations are only possible if donors are clinically diagnosed as "brain dead"—or have a non-survivable injury—and are physically in a hospital with ventilated support.[15]

That's why increasing the reservoir of people who declare themselves to be organ donors is a critical public-health objective and the backdrop to a famous policy nudge: for many years, donation rates were low around the world because people had to specifically request—or "opt in"—to become organ donors and receive a special card (or mark on their driver's license) to that effect. The situation improved dramatically when governments changed their default to "presumed consent"; that is, when governments assume that citizens want to donate their organs after they are gone unless they specifically "opt out" and request *not* to be an organ donor.[16]

The results from this policy shift have been dramatic. The Global Observatory on Donation and Transplantation reports that in 2003, after organ donation became the default option, 99 percent of people in Austria were organ donors as opposed to 12 percent of people in Germany, who still had to actively opt in to donate.[17] Twenty years later, the contrast continues.

The experience of Denmark and Norway, two sociologically comparable Scandinavian countries, is instructive. Denmark still maintains its "opt in" policy for organ donations, while for many years Norway has implemented the opposite default approach, assuming that its population wants to donate organs unless they specifically "opt out." Even though Norway is almost 10 percent less populous than Denmark, between 2010 and 2020 it had 40 percent more kidney transplants from deceased donors.[18] In 2022, a mere eighty-four Danes donated organs posthumously, twenty-one fewer than during the previous year.[19] In short, a simple nudge, unnoticed by most citizens, saved many Norwegian lives. Nudges can encourage all sorts of socially desirable individual behaviors, including many that reduce carbon emissions.

Ultimately, the defining feature of nudges is that they are not coercive: people may face manipulated environments and default settings that subtly lead them to a particular behavior, but they are always able to do something else.

In a recent interview, Cass Sunstein—nudge grandmaster and coauthor of the bestselling book on the subject—shared his threshold criteria for creating effective nudge policies:

> Nudges should be transparent, not covert or hidden. They should be in the interests of the people who are being nudged and consistent with their values. They should be subject to political safeguards, in the sense that if the people don't like them, they should be able to say, "We don't want that one." And they should be consistent with constitutional understandings in the relevant nation. We're very focused on ensuring that nudges are compatible with human dignity. If you're nudged and you think, "That was awful. Why did that happen? I'm sadder and poorer," that's an unethical nudge.[20]

Perhaps because they usually do meet these standards, nudges tend to be popular with the public. A media review of 443 articles reporting on different policy nudges found that coverage was overwhelmingly positive.[21] Nudge theory is increasingly embraced by governments in public policies. The United Kingdom was the first country to establish a governmental, seven-person "Nudge Unit," which evolved into a 200-strong Behavioral Insights Team, now recognized as a global social-purpose company.[22] The European Commission followed suit, even setting up a Competence Centre on Behavioural Insights[23] in order to influence citizens to make decisions consistent with regulatory objectives.[24] And in the United States, the White House Office of Science and Technology Policy established a Social and Behavior Sciences Team in 2014 to translate the insights of behavioral economics into new policy initiatives. Unfortunately, the unit was sacked soon after the Trump administration took over.[25]

In recent years, climate policy has also begun to see an assortment of nudge interventions enacted to gently push the public to adopt climate technologies. The policies focus in particular on promoting individual energy conservation as well as efficiently managing resources and garbage. Not surprisingly, nudge effectiveness, including the duration of beneficial effects, depends on how the policies are designed.

Green Nudges to Reduce Carbon Footprints

Typically, nudge policies in the climate space can be divided into four categories: provision of information; changing the physical environment; creating a "green" default option; and applying social norms and feedback.[26] A review of each category below reveals advantages and disadvantages; not all nudges are created equal.

Providing Information

Most people would rather use less energy to heat their homes or buy appliances that consume less electricity. But very often they don't do anything about it. It is easy to understand why: the gap between instinctive, pro-environmental impulses and actual behavior is often just a function of access to information. The public simply does not know the full implications of its actions or the impacts of the products it buys. Providing relevant information in the hope of influencing consumer preferences constitutes the single most common nudge found in government programs. The goal is to provide information that is both accessible and credible to the public. When presented properly, information can make bewildering decisions, in bewildering areas like energy consumption and efficiency, less complex.

The effectiveness of "reminders" as a nudge that reduces consumption has been repeatedly demonstrated in household water-saving. Households receiving text-message notifications about the importance of conserving water when using showers and appliances saw water usage drop by 23 percent relative to households who were simply given leaflets.[27] During a dry spell in Georgia, text messaging to over 11,000 households led to meaningful water-saving behavior relative to households in the control groups who were not contacted.[28] In the midst of a terrible drought in Brisbane, Australia, the government desperately sought to reduce residents' water consumption. The centerpiece of its campaign encouraged shortening the average length of individual showers from seven to four minutes by distributing over one million free four-minute timers to the public.[29] This less-than-subtle reminder about acute scarcity worked beyond expectations: households cut back on a full fifth of the water they had previously used.

Timing can determine just how effective information-based nudges can be. Tools like in-home smart meters that delay energy usage and remind residents about the extent of their energy consumption (and the costs) have been shown to reduce electricity usage significantly during peak hours. Indeed, in some countries, as part of the installation process, electricity suppliers are required to install smart meters, as well as free in-home displays. When placed prominently in the home, they provide feedback to residents about electricity footprints in real time, thus nudging consumers toward lower demand and thus lower utility bills. Smartphone apps are also used to deliver similar notifications.

A study in the United Kingdom found that homes with in-home displays consumed 1.5 percent less gas and 2.2 percent less electricity than control homes without these systems. This was somewhat below the 3 percent reductions that the researchers had hypothesized. On the whole, smartphone-app information systems appeared to be less effective than in-home displays, but each had its advantages: in-home displays provide ongoing ambient feedback. Because customers have to proactively open their app to see the traffic light–style feedback about their energy consumption, fewer household members end up seeing it. The displays also do a better job of providing "initial learning" about electricity usage. As with the app, customers can examine historical data and set budgets.

On the other hand, apps provide more-detailed energy-saving tips and personalized feedback. In theory, push notifications encouraging people to monitor their electricity and gas expenses can overcome some of an app's disadvantages.[30] The reality, however, is that not everyone finds the time to download an electricity app to their phones.

A nudge can also help the public internalize the extent of its energy use over time. A Hawaiian research group demonstrated that, depending on the time of day, there was an average drop in energy consumption among 11 percent of households that had displays installed in their homes. What made this study design unique was its ability to tease out the reasons for the change in behavior. The researchers tested two logical explanations: either people learned about the relative level of electricity consumed by different appliances or activities and applied this knowledge to change their energy-consumption patterns . . . or alternatively, by constantly reminding people about the electricity they were consuming, energy usage (and the potential to conserve it) simply became a more salient force in their lives.

In the end the results were unequivocal: no meaningful differences were found in either the consumption or the knowledge of the two treatment groups. In other words, long before the end of the experiment, all the Hawaiian households had learned about the differential cost of electricity usage and had internalized the energy implications of different domestic activities.[31]

Labeling of consumer goods, based on efficiency performance, is another information-based nudge. It is especially effective when there is no meaningful price differential between an energy-efficient and an energy-inefficient product. A British study suggests that labeling alone increases selection of energy-efficient products by 5 percent. Encouraging consumers to purchase energy-*efficient* appliances often comes down to getting them to think beyond the immediate, low price of a less efficient product. When factoring in long-term savings, picking low-carbon appliances becomes a "no-brainer." A 2021 nationally representative study of Americans purchasing refrigerators took on this issue. When energy costs became more conspicuous, and people were primed with future-oriented messages, they purchased 24 percent more energy-efficient appliances than a control group that did not receive this information. "Loss-framed" messages, which highlight the opportunity cost of inefficient appliances, reduced the myopia that characterizes consumer decisions when buying refrigerators.[32]

Energy ratings need not be limited to appliances. For instance, assigning energy-efficiency ratings to homes can create a significant profit incentive. In 2008, 25 percent of all houses sold in the Netherlands already had energy ratings. Researchers reported that houses with superior energy ratings sold for a 3.6 percent premium relative to other, comparable homes without distinctive energy ratings.[33] In Denmark, houses built to conserve energy, on average showed a $27,000 advantage at time of sale when their energy-efficiency was made public. This windfall only captures about 50–66 percent of the projected savings that a home rated for high energy-efficiency is expected to deliver over a thirty-year period.[34]

Even before buildings are constructed, it makes sense to use labels to nudge in the direction of reduced embedded-energy building materials. The 2022 American Inflation Reduction Act allocations include $350 million for grants, technical assistance, and tools involving carbon-labeling. The goal is to help manufacturers, institutional buyers, real estate developers, and builders lower the levels of embodied carbon and other greenhouse-gas emissions associated with all stages of production. This includes use and disposal of construction materials and products including steel, concrete, asphalt, and glass.[35]

One of the most contentious areas of climate policy involves the prodigious carbon footprint that meat and other animal-based foods generate. This presents a dilemma. Many people push back from "climate hawks" who pass judgment on people's eating habits and advocate constraints on their diets. On the other hand, emissions from the food system are too significant to ignore.

An Australian and American team considered this sensitive issue, conducting two case–control studies. In the first, participants were asked to estimate the carbon footprint associated with producing and transporting a range of different foods. The results showed the significant extent to which carbon emissions are underestimated by the general public when they are asked to share their impression about the effect of different meals—especially meat—on emissions. The researchers likened the prevailing ignorance to a "blind spot."

In the second experiment, participants were given a list of food items, each characterized by name, serving size, calories, and price. Yet only the "treatment group" received carbon-intensity values. These were translated into units with which people were familiar. The researchers called them "light bulb minutes" or the amount of energy it takes to power a light bulb for a minute. For instance, by one assessment, a serving of beef soup requires energy roughly equivalent to a light bulb turned on for 2,127 minutes—or almost 36 hours![36] The underlying research question was as follows: To what extent could a "carbon label" close information gaps and enable the public to make informed dietary decisions to better understand the climate implications of what it eats?

When the impact of a food's total life cycle was presented clearly by using a familiar reference unit (with emissions translated into the aforementioned light bulb minutes), researchers identified a meaningful drop in consumption of higher emission foods. The researchers concluded that a well-designed carbon label could significantly change the public's patterns of consuming meat and other high carbon foods.[37]

The truth is that there is a debate in the academic literature about the effectiveness of environmental labeling altogether. There are experts who argue that consumers are largely unaffected by green seals and other labeling programs.[38] Many people are already wary, automatically suspicious of disingenuous greenwashing. Other researchers are convinced that it is possible to influence consumption patterns by making accessible messages about products' adverse impacts.[39] The data suggest that they are right.

A critical question in designing an informational nudge involves "duration": Does the effect wane when people stop receiving an ongoing flow of relevant information? To better understand the dynamics, the University of Hawaii team went back to the In-Home Electricity Display nudge. It compared a treatment group of households with in-home displays who were constantly shown their level of electricity consumption to a control group whose electricity was measured but not displayed to residents. The treatment group was then divided into two: half of the families had in-home displays for the full duration of the study; the other group's display was discontinued before the end of the experiment. Contrary to expectations, the discontinued treatment-group subjects appeared to retain the knowledge they had acquired and continued to exhibit energy-efficient behavior. The researchers concluded that learning plays a more prominent role than salience in driving energy conservation.[40]

In order to influence consumer decisions, the way information is framed is critical. For instance, the US Department of Transportation and the Environmental Protection Agency realized that describing a car's fuel efficiency in "miles per gallon" is far less effective for promoting climate-friendly vehicles than characterizing a car's performance according to "gallons per hundred miles." Federal reporting rules were modified accordingly. In another case, American officials realized that telling homeowners living in 100-year floodplains that they face a "1 in a 100 chance of flooding" during any given year leads to complacency. The government now lets them know that there is "a 1 in 4 chance of suffering a flood during the course of their 30-year mortgage."[41]

Information so poorly scripted that it becomes annoying can produce a "boomerang" effect. Some people are contrarian, responding by overconsuming or showing their displeasure by opting for the opposite, anti-environmental behavior. For instance, an electric utility sent letters to consumers comparing their energy consumption to that of their neighbors. Republicans, presumably less enamored of conservation, responded to the campaign by increasing their electricity consumption—as opposed to affiliated Democrats, who lowered theirs. Even worse: anti-litter campaigns, whose messaging in retrospect was sloppily phrased, may have actually increased littering by highlighting its prevalence as a socially acceptable behavior.[42]

Changing the Physical Environment

Food waste is a significant source of global emissions. According to the United Nations Environmental Program, it accounts for 8–10 percent of the total carbon released worldwide.[43] As the population grows and electricity becomes increasingly decarbonized, this proportion will grow. In developing countries, roughly a third of the crops raised is lost due to inadequate refrigeration and infrastructure, along with lack of information about markets. In most of the developed world, however, food waste is all about human behavior and poor planning. One reason why so much food is thrown out is that portions are impossibly large, leaving sizable quantities of excess uneaten food. The problem of wasted food is never more acute than in "all-you-can-eat" buffets, where people's eyes invariably are "bigger than their stomachs" and there is no economic downside to taking too much.

One way to address this in restaurants is to make plate size smaller. Researchers hypothesized a domino effect where smaller plates would lead to smaller portions, resulting in reduced food waste. This nudge was tested in fifty-two Norwegian hotels, seven of which served as control groups. After plate size was reduced from 24 to 21 centimeters, food waste also dropped by 19.5 percent relative to the control group. In another experiment, a social cue was added where guests were reminded that the buffet could be visited several times—implying that they didn't have to pile on all the alluring delicacies at once. When this element was added to the plate-size intervention, food waste dropped an additional 20.5 percent. All told, the calculations suggest that for every centimeter reduction in plate size, there is a 7.4 percent drop in food waste.[44]

Restaurants can reduce meat consumption simply by changing the location of the meat entrees at cafeterias or on menus, while at the same time making vegan and vegetarian options more conspicuous and appealing. In one study at a Midwestern American university dining hall, the odds of students choosing vegetarian meals significantly increased when non-meat items were made more conspicuous. Enhanced vegetarian choice architecture eliminated the need for sanctimonious messaging or economic incentives.[45] The physical environment can also be designed to increase recycling and reuse. Examples of such nudges include the layout of waste bins, offering curbside pickup for recyclable materials, or simply adding accessible recycling receptacles alongside trash cans.

Creating Climate-Friendly Default Options

The power of inertia makes default options a particularly effective nudge. Default options do not necessarily exploit individual laziness, procrastination, fear, or distraction—although these powerful forces surely play a role. Still, something far deeper is involved. In his research, Professor Kahneman documented an *endowment effect*, or what is also known as a "loss aversion" bias. This means that most humans have an irrational tendency to hold onto things they already own and to preserve the status quo. People exhibit a natural dislike for losing something, even if there is no reason for a sentimental attachment. Kahneman gives the example of a colleague who collected bottles of wine whose value never exceeded $100. When offered $1,000 dollars for a bottle, he was loath to sell it, even though the economically rational response meant happily accepting the offer.[46]

When faced with a "default," many people also infer that a higher authority, endowed with superior wisdom, approved the decision. This makes it easier to simply go along with the nudge. Cass Susstein and Lucia Reisch explain:

> Choosers may believe that they have been given an implicit recommendation (perhaps from a private institution, perhaps from public officials), and that they should not reject it unless they have reliable private information that would justify a change. If the default choice is green energy, it is tempting to think that experts, or sensible people, believe that this is the right course of action. . . . Especially if they lack

experience or expertise and/or if the product is highly complex and rarely purchased, they might simply defer to what has been chosen for them.[47]

As the world's grids become increasingly reliant on weather-dependent, intermittent electricity, individual energy-consumption patterns become critical. This explains the growing interest in electricity-consumption nudges. The same phenomenon that led consumers to stick with green electricity in Schönau is now taking place on a larger scale.

Another energy utility in the south of Germany, the Engergidenst GmBH, changed the default for its 150,000 private and business customers to include a basic green tariff for cleaner energy. If they took the trouble to consider them at all, customers were offered three alternatives. The default "green tariff" was based on electricity with a meaningful fraction of renewable sources supplying it. Consumers who preferred the green tariff were not required to respond. The second alternative was a 23 percent more expensive package, involving an even higher percentage of renewable electricity than the default mix; the third option was a less expensive, more polluting, *gray* alternative source that was 8 percent cheaper than the default. Again, 94 percent of the customers predictably chose *not* to "opt out" and stayed with the default green tariff.[48]

German economists replicated these findings in other regions: the default option, which presented a relatively expensive, renewable electricity package, always succeeded in contributing to energy decarbonizing. This happened without any of the pushback associated with mandated requirements. Renewable defaults are not limited to Germany. The city of San Jose, California, gave its residents a chance to go back to a conventional Pacific Gas & Electricity package if they wanted to opt out of the city's "GreenSource" default.[49] The results were essentially identical. People stuck with the cleaner option.

The German research suggests that when green energy becomes the default choice, even if it is more expensive, purchases of renewables increase tenfold. Behavioral economists tried to better understand the motivational underpinnings of such consistent energy preferences. They rejected the notion that consumers tend to accept a default renewable energy option simply because they don't read or don't understand when they do read, remaining unaware of the implications. Rather, researchers hypothesize that opting for a green default was a conscious decision.[50] An endowment effect clearly emerges from their survey: on average, participants reported a willingness to pay to switch to green electricity of €6.59. When asked how much they would have to be paid to give up green energy in favor of a conventional source, the figure was double: €13. These findings are highly consistent with differences found in other *contingent value* studies that distinguish between "willingness to pay" and "willingness to accept."[51]

Defaults emerge as a significant force in other energy contexts. For example, back when incandescent light bulbs were the default form of lighting, a Columbia University study found that they were preferred 43.8 percent of the time. But when the far more efficient compact fluorescents became the default, the percentage opting for energy-intensive incandescents dropped to 20.2 percent.[52] Or there are motion detectors that automatically turn lights off when people are not moving or not present in rooms. These constitute an increasingly common default in hotels. Or screen savers that appear routinely on computer screens and other programmed mechanisms for turning off machines and appliances when they are not being used. These can all be modified; very few people do, however.

Particularly germane to today's electricity-transmission challenges is the public's attitude to installing smart grids in homes. Given its intermittency, renewable electricity will hit constraints far earlier if a utility can't remotely control some of households' electricity consumption. In other words, it is important to convince the public to cede some of its autonomy over electricity consumption in order to optimally balance the grid. When researchers explored this challenge, once again the willingness of the public to accept an imposed smart-grid system in their homes depended on the default context: hypothetically, acceptance went up when people knew they had the power to opt out in order to cancel automatic external control.[53] As smart grids become more commonplace, defaults will be important in all phases of domestic life: from the optimal time to run the clothes dryer to when to charge electric cars.

Examples of nudges saving resources are everywhere: Scandinavian economists Johan Egebark and Mathias Ekström found a default effect for computer-printing behavior: when the default printer setting was changed from single-sided to double-sided, the consumption of printer paper dropped by about 15 percent. There was no evidence of a decline in this effect 28 weeks after the change.[54] This green "default" is a favorite of green campus initiatives at universities and schools around the world because it is so easy, economical, and logical. As Internet access became largely universal over the past decade, many companies went paperless; electric companies, water and gas utilities, credit card firms, tax agencies, and banks have all moved to make their billing take place online—unless a consumer specifically asks to receive a bill in the mail. But this rarely happens.

Setting the default at a level that does not push people beyond their comfort zone turns out to be tactically important. At some point, the majority *will* opt out of an excessively uncomfortable environment. One study conducted at the

OECD offices in France during the winter considered the effect of reducing the default temperature on energy use. After a small "default" reduction in temperature—by 1 degree Celsius—employees apparently did not perceive the difference to be sufficiently significant to return thermostats to the original level. But when the default temperature dropped 2 and 3 degrees, they were motivated to take action and return the heat to the original level. This eliminated the potential energy-savings benefit that had been gained from a more modest reduction. The researchers conclude the following:

> If the reduction in default temperature is too large, then occupants respond actively, increase their temperature settings, overriding the effects of the change in the default setting. In quantitative terms, our results indicate that a reduction of the default temperature from 20°C to 19°C would decrease energy use, but a reduction to 17°C would have no effect.[55]

Conveying Social Norms and Feedback

People are often unaware that their behavior is unsustainable. One way to convey this unwelcome news, along with an implicit suggestion for improvement—without being excessively smug or obnoxious—is by describing the normative behavior of a relevant group of peers. Perhaps the most successful anti-litter campaign in history was achieved by appealing to the collective identity (and pride) of Texas residents. The celebrated "Don't Mess with Texas" initiative broadcast the nasty scourge of trash in public spaces, leading to a reported 72 percent reduction in the amount of litter across the state.[56] The implicit message to litterers was that they were not meeting the basic standards for good citizenship expected by their fellow Texans.

Nudges designed to reduce energy usage can send a similar message. Many utilities forward monthly reports that compare individuals' consumption with their own usage in the past—or with that of comparable households. Bar graphs that contrast electricity usage with that of more-efficient neighbors during comparable months or years hit home. Following this with concrete suggestions for improving energy-efficiency can create meaningful change. Such a nudge was evaluated in Sacramento, California, and sites in Oregon, with statistically significant, albeit modest, results: reductions in energy consumption of 2 percent.[57]

Beyond their role in changing *individual* consumer behavior, nudges have proven highly effective in *institutional* settings. An initiative by the British government set a target of reducing carbon emissions in its departments by 10 percent within a year's time. This was an ambitious project, involving 3,000 buildings and 300,000 civil servants. The program started by dropping the default settings of heating and lighting systems: buildings were never overheated or overcooled, with temperatures staying within a thermal comfort range of 19–24°C. Real-time displays of energy use were installed in buildings, and monthly performance reports were circulated detailing how each agency was doing relative to its "competitors." The project's final report explained: "This introduced a competitive element to departmental performance and a strong incentive for departments to avoid the reputational loss of being seen to perform poorly on the Prime Minister's commitment." After a year, calculations suggested that the reduction in energy usage actually exceeded 10 percent.[58]

A Scandinavian team compared studies on the effectiveness of different energy conservation nudges and offered a summary of the most promising interventions. Their report showed universal success in the use of green defaults that deliver green energy: 90 percent of consumers worldwide retained defaults that promote renewables. Labeling, on average, increased adoption of an energy-efficient product by 440 percent.[59]

The Ethics of Nudges

Advocates of nudges often emphasize the ethical advantages of "libertarian paternalism" and its ability to encourage the public to pursue socially optimal behaviors without infringing on individual freedom. Yet there is no shortage of critics, even when policies merely seek to steer citizens in a given direction without resorting to domineering regulations. According to detractors, there is an arrogance behind the very notion that policymakers know what is best for the public, in all cases—not to mention what people actually want. This critique is not new. Indeed, one of the great moral philosophers of all time, John Stuart Mill, raised a similar concern in his essay, *On Liberty*. Mill's essay is circumspect about governmental overreach:

> The interference of society to overrule [a person's] judgement and purposes in what only regards himself, must be grounded on general presumptions: which may be altogether wrong and, even if right, are as likely as not to be misapplied to individual cases. . . . [60]

In other words, even if policy makers can legitimately identify socially optimal behaviors, their ability to translate these assumptions into defaults or messaging interventions is highly imperfect.

John Stuart Mill was a very brilliant man. But his position on these matters appears a little naïve. The truth is that people are constantly being influenced by both arbitrary and intentionally designed choice architecture. Cass Sunstein, probably the most effective purveyor of the *Nudge* vision, explains that subtle— or even not so subtle—signals are everywhere. "Neutral choice architecture" is an illusion. People inevitably are affected by the way choices are presented to them. When policymakers adopt a nudge, they simply reorganize the many signals to which the public is already exposed, but do so in a more edifying way. Ignoring choice architecture's existence and its effect on behavior makes little sense and can even be irresponsible, leaving authorities complicit in creating messages that are at odds with the public good.[61]

Nudge critics are on their most solid ground when they question policies that emphasize broader societal goals over the well-being of the individual citizen. Another concern is that people from lower socio-economic levels with less education may be less inclined to opt out of defaults, relative to wealthier or more-educated individuals.[62] If a policy is designed for collective well-being at the expense of the individual "nudgee," it may perpetuate inequalities. For example, a nudge that favors expensive electric vehicles may be regressive and thus increase inequality. In such cases, increased transparency can remedy those concerns, even if it may reduce the policy's overall effectiveness. If the purpose of the nudge involves a societal goal, it is well to clearly state this up front, in a way that the public can understand, before setting a nudge in motion.[63]

Excessive reliance on "nudge policy" runs the risk of crowding out support for other, more ambitious, climate policies. A team from Carnegie Mellon University effectively demonstrated that nudges aimed at reducing carbon emissions can have a pernicious *indirect* effect. By offering the false promise of a "quick fix," a nudge might undermine support for bolder climate policies capable of far greater impact. The CMU study suggests that support for carbon taxes dropped when respondents were given the alternative of less-invasive nudges. The researchers conclude that nudges decrease support for more-demanding policies by providing false hope that problems can be tackled without imposing meaningful costs. They suggest that by disclosing the relatively small impact of nudges, policymakers can eliminate the "crowding out" effect while retaining support for the nudge as a supplementary intervention.[64]

Ultimately, confronting climate change will require long-term modification of human behavior. Whether or not nudges can produce lasting behavioral change and prevent backslide is still unknown. For example, without a constant reminder to save electricity or eat less meat, will people still prefer the low-carbon alternative, based on an internalization of its importance, or even simple inertia? A few experiments suggest that the effect of nudges can be persistent.

One of the great tactical advantages of using nudges over other policy instruments is that government agencies—and corporations—usually do not need statutory authorization to adopt a climate-friendly nudge. Despite its significant influence, adjusting choice architecture predominantly remains a discretionary administrative decision, one that can be nimbly applied. From a practical perspective, this is a good thing.

Research confirms time and again that nudges can serve a very useful role in encouraging an individual's climate-friendly behavior. Nudges can also be readily deployed to push corporations toward new environmental norms without inciting them to push back by mounting legal challenges on the grounds of "arbitrary and capricious enforcement actions." For example, if a corporate "climate index" became a salient performance metric for investors, such a nudge could resonate through boardrooms worldwide.

Even without mandatory proscriptions, simply reporting carbon emissions can induce behavioral changes. It will not be long before this policy becomes more than just a hypothetical musing. In March 2022, the US Securities and Exchange Commission published a 500-page proposal that would require publicly traded companies to disclose a range of information, including their greenhouse-gas emissions footprints, emission-reduction goals, and plans to achieve them including the use of carbon offsets.[65] Presumably, companies will need to report how their board of directors governs climate risks and how climate change is expected to affect line budget items.[66] Felix and Milica Mormann from Texas A&M University hypothesize that adding a climate rating to corporations' reporting rules would lead to a more than 50 percent increase in investment in the stock of climate-conscientious companies.[67]

Can nudges help to accelerate innovation? Undoubtedly, choice architecture can be manipulated in research programs and applied in "calls for proposals" to push scientists and companies toward priority areas. This might make a modest contribution to climate tech R&D programs. (Simply offering more-generous grant funding would probably be a more promising strategy.) Nudges' real potential influence, though, is through creating conditions that encourage people to adopt new, cost-effective, low-carbon technologies. Climate tech innovation is not just about inventing new gadgets. It also means scaling new technologies once they have a proven ability to reduce carbon emissions. This involves billions of individual decisions. Nudges have an important role to play.

CHAPTER 9

Disruption

Professor Mark Jacobson, an atmospheric science researcher, emerged fifteen years ago as the chief scientific proponent of a net-zero future. For most of his thirty-year career at Stanford University, Jacobson developed three-dimensional computer models to better understand the interplay between air pollution and the weather. His computer code provided the analytical infrastructure for widespread further research.[1] Early on, Jacobson became very concerned about the health implications of soot and black carbon emissions. In 2007, he appeared before Congress to testify that the existing US standards were insufficiently stringent.[2] As time went on, the problem of climate change became increasingly central to his thinking and research agenda.

Comfortable with modeling on a large scale, Jacobson began to tinker. Relying on existing climate technologies, he explored different greenhouse-gas mitigation scenarios on a global level. These initial forays into decarbonization pathways took place long before renewable energy prices began their slide into freefall. Nonetheless, as he began to crunch the numbers, Jacobson became firmly convinced that a full transition to a non-carbon-based energy system was completely feasible.

Publishing with UC Davis colleague Mark Delucchi in *Scientific American*, Jacobson called for the installation of 3.8 million large wind turbines, 90,000 solar plants, and numerous geothermal, tidal, and rooftop photovoltaic installations worldwide. Nuclear power was not part of his "net-zero" vision. This was not only because of the dangers associated with radioactive waste disposal and terrorism risks. It was mainly a question of efficacy. When reactor construction as well as uranium refining and transport are all included in the equation, Jacobson calculated that nuclear power causes twenty-five times more carbon emissions than wind turbines, making it a cure worse than the disease.

Departing from the usual dry, analytical scientific style, Jacobson and Delucchi's article, entitled "A Plan to Power 100 Percent of the Planet with Renewables," waxed visionary about their strategy:

> Our plan calls for millions of wind turbines, water machines, and solar installations. The numbers are large, but the scale is not an insurmountable hurdle; society has achieved massive transformations before. During World War II, the US retooled automobile factories to produce 300,000 aircraft, and other countries produced 486,000 more. In 1956 the US began building the Interstate Highway System, which after 35 years extended for 47,000 miles, changing commerce and society.[3]

Subsequent studies, reports, and even a highly regarded book[4] provided far more detailed scenarios. First, he drew up blueprints for all the US states and then went on to propose sustainable pathways for a hundred countries.

Jacobson's bold net-zero vision was initially questioned and occasionally derided as unrealistic, unworkable, or even delusional. Undeterred, he continued to model, report, and publish additional well-documented articles, expanding his vision of possible routes to net zero. As time went on, more and more scientists became convinced that the net-zero-emissions route not only was possible—but was in fact—the only way to save the planet.[5] By 2016, the *Proceedings of the National Academy of Science* (PNAS), among the world's most prestigious scientific journals, awarded the Cozzarelli Prize, which recognizes particularly outstanding contributions to science, to Jacobson's article "Low-cost solution to the grid reliability problem with 100 percent penetration of intermittent wind, water, and solar for all purposes."[6]

Eventually, world leaders came around and began to make commitments to reach net-zero-emissions profiles by 2050. In 2023, the UN reported that 140 countries, including the world's biggest carbon polluters—China, the United States, and the European Union—had committed to achieving net-zero targets, covering about 88 percent of the planet's total emissions.[7]

But what of the occupational implications of such a dramatic and rapid redesign of the world's energy systems? In a major 2017 analysis, Jacobson and his coauthors argued that a low-carbon economy would be good for global employment.[8] Actually, *very* good. They calculated that transitioning to a net-zero economy would create 24.3 million more permanent full-time jobs than would be lost due to decarbonization.[9] The calculations include both *direct* jobs for onsite construction, operation, and maintenance of electricity-generating facilities, and also *indirect* jobs associated with the revenues produced by renewable supply chains. These involve any employment associated with related construction materials and component suppliers, as well as induced jobs from spending and reinvestment of earnings. One can argue whether the jobs associated with increased business at local restaurants, hotels, retail stores, or even childcare providers should be counted in the overall equation. (Changes in employment due to possible changes in energy prices were not included.) But even disregarding Jacobson's generous assumptions, the evidence is compelling that decarbonizing the economy will create a significant number of jobs.

This is not to say that dramatic transitions are without repercussions and socio-economic consequences. The history of transformational technologies is replete with "winners"—and "losers." There will be individuals, communities, and even countries that pay a price as the world races toward a low-carbon economy; many already have. The challenge of meeting their needs has become known as a "just transition" to a low-carbon economy. This chapter considers what can be done to that end.

The Employment Dividend of a Net-Zero-Emissions Economy

Looking ahead, a pair of Greek economics professors, Stella Karagianni and Maria Pempetzoglou, paint a very grim picture of a 2050 net-zero world:

> Rapid change in the energy model, without being accompanied by supportive policy measures for the affected areas and their populations, is bound to degrade economic development and destroy jobs in the areas where power generation stations are located, strengthen migration flows, increase energy prices and intensify energy poverty, deteriorate trade balances, and enhance environmental degradation. All these factors will contribute to widening the gap of income inequality among households.[10]

This makes for powerful rhetoric. But there is little solid data to validate such alarming scenarios. Indeed, the vast majority of studies conducted on this topic suggest an entirely different future: many more jobs will be gained by moving to a more sustainable global equilibrium than will be lost. Many of them will be good jobs. In the past, there have been concerns about new technologies disrupting the livelihoods of people who rely on an old technology that is on its way out. Frequently, the alarm was unfounded. For instance, during the nineteenth century, bookshops saw free public libraries as an existential threat to their businesses. But ultimately, the new libraries only increased people's appreciation of books and expanded demand. Years later, Hollywood producers warned that video rentals would be a menace to the movie industry. But they quickly stopped complaining when they saw how much it contributed to expanded distribution and to their revenues.[11]

For many hapless occupations, however, apprehensions did prove prescient. Introduced in the 1980s, desktop publishing took less than a decade to make typesetters completely superfluous in the printing business, eliminating a profession that had employed at least 10,000 people.[12] A few hundred thousand American women worked as switchboard operators in the 1930s, but as soon as it became possible to dial other phones directly without the assistance of a switchboard operator, their ranks began to dwindle. In 2021, the Bureau of Labor Statistics reported a mere 5,000 people working as "telephone operators."[13]

Travel agents, lamplighters, film projectionists, ice cutters, telegraph operators, milkmen, elevator operators—these are just a few on a long list of honorable vocations that all but disappeared through no fault of their own. They were simply on the wrong side of technological history. While, overall, the world's economy and overall job market grew with each leap of progress, at the micro-level there were people who paid a price. Not every stable hand was able to enroll in a community college and emerge an auto mechanic. That's why it is wrong to be dismissive about the socio-economic implications of a decarbonized economy.

Whether for the sake of compassion or political expediency, it is well to consider an appropriate policy response to the disruption and displacement resulting from a new low-carbon labor market. At the same time, if the world does not act to mitigate climate change, the consequences for workers and livelihoods around the world will be far more egregious than any adverse effects caused by the transition to a sustainable economy. The subsequent section considers both the anticipated losses and gains for the world's labor market as the world moves toward a new net-zero reality. It then introduces four ways to address the negative distributional impacts of decarbonization.

Creative Destruction

No historic discussion of technological innovation and disruption is complete without some reference to the inimitable Joseph Schumpeter, the first economist to devote significant attention to the topic. Professor Schumpeter actually gets credit for inventing the term *entrepreneur*: "an innovator who implements change in an economy by introducing new goods or new methods of production." One of the rare giants of economic scholarship who had actual government experience, Schumpeter served as finance minister of Austria in 1919 and later ran a bank. A particularly colorful and intelligent man, he was as openly proud of his horsemanship and amorous capabilities as he was of his academic brilliance. And brilliant he was. As a Catholic, he found that the Nazi presence in the 1930s made life in Europe intolerable. Most of his theorizing and prolific writing, therefore, actually took place at Harvard University. For regular readers of *The Economist*, Schumpeter is now immortalized by a regular column that bears his name. Since 2009, "Schumpeter" delights readers with encouraging news of innovation and economic progress.

Above all, Joseph Schumpeter was an optimist. His British "rival," John Keynes, was more of a pessimist who tended to focus on steadily declining marginal returns to capital. Joseph Schumpeter, on the other hand, believed that it was not economic forces or "workers of the world" who drove economic advancement. Rather, *innovation* was the engine for meaningful societal and economic evolution. His observations convinced him that business cycles are fueled by dramatic technological advances that come in pulses and catapult economies forward.[14]

Schumpeter was well aware of adverse consequences that technological disruption might cause workers, professions, and businesses. He also realized that monopolies might emerge (albeit briefly) after introducing a particularly unique product. But he saw this as collateral damage, unavoidable and perfectly acceptable in the higher mission of improving the general standard of living and the collective quality of life. Indeed, he unabashedly celebrated what became his signature slogan: "creative destruction." An artifact of a different era, his disregard for laborers whose jobs are usurped by a new technology may appear obtuse today. But his basic point is compelling: real progress may leave a few people behind, but the vast majority of humanity will be ahead. A just transition to a sustainable economy requires societies to minimize the number of people consigned to obsolescence and to help communities that are disproportionately affected.

Global Warming and the Global Labor Market

The International Labor Organization is a United Nations Agency, founded in 1919 under the now defunct League of Nations. Its mandate is to advance social and economic justice by setting international labor standards.[15] In 2019, while it was celebrating its centennial anniversary, it released a chilling report about the ramifications of the climate crisis for laborers.

Today, most people work indoors, especially in developed countries. The 2019 report reminded the world there are many people who don't: this includes farmers, construction workers, refuse collectors, emergency repair workers, tour guides, park rangers, law enforcement officers, and athletes—to name just a few. Indoor workers may not be entirely sheltered from the consequences of global warming, either. Buildings without proper ventilation or air-conditioning can become even more stifling than the outdoor environment. For billions of people, excessive heat exacerbates occupational risks and can harm physical and psychological health. At 24°C (75°F), labor productivity begins to fall. Once temperatures reach 33°C (or 91°F), "a worker operating at moderate work intensity loses 50 percent of his or her work capacity."[16]

Of course, climate change affects different regions in different ways. By 1995, for example, countries in southern Asia had lost an estimated 4 percent of their total working hours (the equivalent of 19 million full-time jobs) as a result of high temperatures. But with the projected warming, and assuming only a 1.5-degree average increase in 2030 temperatures occurring worldwide, the UN report projects that 5.3 percent of total working hours worldwide will be lost due to heat extremes. That is the equivalent of 43 million full-time jobs. Economic repercussions will be worst in regions that are already warm, like Central America or the American Southeast and Southwest. Among the economic sectors that will be most affected are the wine industry, tourism, farming, fisheries, energy, and food/beverage.[17]

All this means that if the world doesn't move expeditiously to stabilize greenhouse gases, many important professions will suffer significantly. Snow sports alone are projected to lose $20 billion over the coming years, and thousands of ski instructors will be out of work.[18] Ocean acidification and other climate-related marine ailments affect fisheries and the people who rely on them for a living. For example, between 1996 and 2017, New England's coastline saw an average drop of 16 percent in fishing jobs—with most of the decline attributed to global warming.[19] Some scenarios suggest that by 2050, climate change could cause a $41-billion loss in sales from global fisheries—which means that many more fishermen will be out of work.

Climate determines one of the two most critical factors in wine-making: the soil–climate equilibrium. When temperatures exceed 30°C (90°F), photosynthesis is actually reduced due to increased respiration.[20] In recent years, frequent floods, along with longer and fiercer heatwaves in wine-growing regions, have damaged grape crops. Wildfire smoke has ruined grapes growing as far as 100 miles away, with wine made from them becoming thoroughly "undrinkable." The news agency Reuters reports that climate impacts in 2022 were responsible for the smallest grape harvest in France since 1957, corresponding to a $2-billion drop in wine sales and thousands of related jobs.[21]

The obvious common denominator of these trends is that climate change bodes poorly for people who make a living outdoors. Ameliorating the impacts on these highly vulnerable workers is yet another reason why the world needs to mitigate climate change and accelerate the transition to a sustainable economy. At the same time, it is wrong to forget the many professions and people who stand to suffer as the world's economy leaves its old carbon-intensive ways behind.

Decarbonization and Disruption

When considering employment trends, it becomes clear that prioritizing renewables is simply hastening the end of an ongoing historic process. The number of coal miners has long since plummeted into near-oblivion. It peaked at 863,000 in 1923, when there were 111 million Americans. One hundred years later, with 334 million US citizens, the number has fallen to a very modest 66,000. In Poland, while there were 400,000 gainfully employed coal miners in 1990,[22] there are only 75,000 working in what was once the country's most critical economic sector.[23] During the period of peak employment, over 50,000 Americans lost their lives in coal-related accidents.[24] Many times more died of black lung disease after years of breathing in toxic dust at their workplace. Even today, exposure to coal particles is associated with 1,000 annual deaths on average among former coal miners.[25] With all due respect to socio-economic disruption, decarbonization promises to be a public-health savior for coal-mining communities.

Coal production is hardly the most affected sector as the economy moves away from carbon. The American Petroleum Institute, a trade organization, reports 145,000 fueling stations across the country. Roughly 127,000 of these stations can be characterized as convenience stores selling gasoline. Most are independently owned and operated. Indeed, only 5 percent of American gas stations are actually owned by oil companies.[26] In 1970, more than 200,000 stations in America were pumping gas.[27] Independent reports estimate that the actual number of gas stations may now be as low as 115,000.[28] There is little question that the number of filling stations in the United States is declining—fast.

There are many reasons for this trend: improved fuel-efficiency, the emergence of electric vehicles, and the preference of young people to travel via ride-sharing and public transportation. Here, too, a shift to a fleet powered by clean electricity will deliver meaningful occupational health benefits. Elevated stomach cancer rates among gas station attendants, along with kidney and liver damage due to exposures to benzene and other volatile organic compounds, may not reach the dimensions of coal miner morbidity and mortality. Nonetheless, pumping gas definitely has its occupational hazards.[29]

Dramatic changes are transforming the world's auto industry. The average life expectancy of a gasoline-run vehicle is eleven years, so the final demise of internal combustion on the roads will take a few decades. Gradually gas stations will begin to disappear altogether. Some may yet be able to pull off a makeover and join the new charging infrastructure. But gas stations are certainly not a growth industry.

Getting to net zero will affect far more people than those whose livelihoods are dependent on the old fossil-fuel-based economy. With the food system responsible for 34 percent of global emissions,[30] phasing down animal-based calories will have ramifications. For instance, with the production of cultured milk progressing apace, dairy farming may soon become a thing of the past. There has already been a dramatic consolidation of dairy herds, with the number of American dairy farms falling by more than half between 1997 and 2017.[31] Unlike coal and gas stations, which have been in decline for some time, overall milk production has actually expanded as of late in the United States, with demand largely driven by dairy exports.[32] The number of cows per farm increased by 139 percent—with 70 percent of milk presently produced on farms with more than 500 cows. This trend is unlikely to continue. Even the most-efficient "megadairies" will be far more costly and environmentally destructive than their cultured competitors.

Cultured milk producers are already making cheeses that are identical to conventional ones. They are much better for the planet: cultured milk uses but 1 percent of the land, 4 percent of the feedstock, and 10 percent of the water—and emits 97 percent less greenhouse gases—than milk that comes from cows.[33] And no animals suffer at all. Even before regulators monetize beef and milk's carbon footprint, dairy farmers will simply be unable to compete economically with the low-cost, high-quality, hormone-, cholesterol-, and lactose-free dairy products that miraculously arise from massive fermentation tanks.

Milk has a head start. It will not be long before cultured meat becomes indistinguishable from that of slaughtered animals. It will be cheap enough, clean enough, and ethically superior enough to completely disrupt the cattle, poultry, and pork industries. The US Food and Drug Administration and the Department of Agriculture have already approved lab-grown chicken for sales across America.[34] These sectors provide livelihoods to millions of people worldwide. Very soon many of them will need to find alternative employment. Luckily, there should be no shortage of new jobs waiting in the decarbonized economy.

The New Decarbonized Job Market

Today, Mark Jacobson's predictions about the low-carbon economic transition's occupational benefits have been validated by a range of authorities. IRENA, the UN's International Renewable Energy Agency, calculates that already 12.7 million new jobs have been created within the renewable energy space. More than a third of these involve the solar electricity sector, in which employment continues to grow at a rate of 6 percent a year.[35] That implies a doubling every twelve years.

Other assessments concur: according to the US Bureau of Labor Statistics, of the twenty occupations likely to see the fastest growth over the coming decade, wind turbine technicians are ranked second, with solar installers in eighteenth place. Neither position pays particularly well, though. The average salary of turbine and solar technicians in the United States is only $50,000 a year, well below the median income and about 40 percent below the average pay for nurses.[36] The growing demand for solar and wind power may change present salary dynamics, however. The International Energy Agency, which represents thirty-one member countries that together are responsible for 75 percent of global energy demand, is even more optimistic. According to its 2022 report, *World Energy Employment*, if humanity reaches net-zero emissions by 2050, 14 million new jobs will be created by 2030 to manufacture, install, and maintain renewable energy facilities. An additional 16 million workers will shift to new roles related to clean energy.

The OECD posits that most carbon-intensive industries are *not* labor-intensive and employ few workers, relative to their economic output. That is why the OECD believes that the low-carbon transition will put many more people to work than those whose livelihoods may disappear.[37] It is true that installing photovoltaic panels on roofs is extremely arduous and not particularly lucrative work. But around 60 percent of the positions created by the new energy economy require postsecondary training and do not involve manual labor.[38] Many of these positions pay well. A 2023 report by the World Economic Forum reports that global demand for "green skills" grew some 40 percent in seven years.

In the United States, the first projections are appearing for the employment impact of the 2022 Inflation Reduction Act. Researchers at the University of Massachusetts predict that the law will be instrumental in creating 9.1 million new jobs during the ten years following its enactment. Over half of these positions are associated with electricity production. Other areas of the decarbonized economy will create jobs as well. *The Economist* suggests that "new jobs in green hydrogen or electric vehicle assembly will probably not constitute a net gain but have the potential to replace workers in oil refining or existing car factories, who will have to be laid off as a result of the new green economy."[39]

Given years of stagnation, the International Energy Agency estimates that investment in electricity grids needs to nearly double by 2030 to over $600 billion per year in order to support global electrification and meet national climate targets.[40] The truth is that the job market is unable to keep up with the mounting demand for professionals who can rewire and optimize the grid. A mere 13 percent of the present labor workforce actually have the skills that organizations need and want.[41] Kate Gordon, an energy policy expert who served as senior advisor to Biden's secretary of energy, is forthright in her assessment of the skill-set gap:

> Twenty or thirty years ago, we convinced everybody going into engineering that they should instead go into computer science. And so we ended up with an enormous amount of people in computer science and very few engineers. Today we find ourselves with this huge national lack of power system and nuclear engineers, both of whom are needed for just about everything right now.[42]

The American labor market really does have a shortage of the professions so critical to creating and scaling climate innovation. US Secretary of Commerce Gina Raimondo has also been open about the mismatch between the technical challenges and available skill sets:

> Every company you talk to, literally everyone, will tell you the thing they're most worried about is talent. The rate limiting factor to growth is talent. And by the way, it's at every level, right? From process engineers, technicians, to the PhD [in] physics and the like.[43]

The US Bureau of Labor Statistics projects that during the years between 2016 and 2026, the country will have suffered a shortfall of six million engineers

. . . or more.[44] The phenomenon appears to be global: Germany faces a shortage of 216,000 electricians to build out its renewable electricity sector.[45] This suggests that there is a silver lining in the occupational turbulence caused by climate tech and innovative technologies. Creating the capacity to fill these jobs constitutes a huge opportunity for labor policy programs.

Transitional Assistance Policies

Modern welfare states offer many different forms of social safety nets for their citizens to catch them should they fall economically or face unusually trying social circumstances or health challenges. In most welfare states there is already a reasonable menu of programs to help people who find themselves unemployed due to the usual vacillations of modern economies and business cycles. There are two reasons why individuals displaced by the transition to a low-carbon economy are different:

1) Changes in the labor market will be happening more quickly in many sectors than under usual circumstances. This makes rapidly adapting to a new, decarbonized reality significantly more daunting.
2) Governments around the world are adopting aggressive policies to expedite deployment of climate tech. When a job is lost as a result of a government-initiated program to phase out entire sectors in favor of alternative technologies, society has a greater responsibility to help the "displaced" find alternative livelihoods.

Whom to assist and how to support them is in no way self-evident. Social science literature has begun to consider where social disruption is most likely to occur from the energy transition and what can be done to rectify it. In 2020, the Stockholm Environment Institute published one of the first such analyses.[46] It relies on historic cases of economic disruptions and distills the lessons that have been learned down to seven principles that should guide policymakers. The authors argue that governments should prioritize regions with low capacity or high dependence on carbon-intensive economic activities.

The seven-point strategy starts by emphasizing that *delaying* decarbonization is fundamentally unjust. Expediting a swift decline in global emissions is essential in order to avoid exacerbating the negative distributional impacts of global warming. Second, a just transition should avoid "carbon lock-in": ongoing support for carbon-intensive industries needs to stop. So does reinforcement of any societal dependence on them. The third point is that when investing in labor diversification, governments need to prioritize support for regions with lower financial capacity, as well as those who bear lower historical responsibility for global emissions. The Stockholm report then calls for direct support for workers, their families, and communities that are disproportionately affected by closures or downscaling as a result of technological transition.

In practice, this means providing funds for training or support, enabling communities to develop new livelihoods. If re-employment is not possible, flexible solutions should be available. The report also envisions *ecological rehabilitation* as part of the energy transition strategy. Any environmental damage left in the wake of the old carbon-intensive economy needs to be cleaned up; related costs should *not* be transferred from the responsible actors in the private sector to those in the public sector. The final two principles reflect the kind of economic ideology that one would expect from a progressive Swedish think tank: it calls for social equity and empowerment of vulnerable social groups to be an explicit policy goal as well as a priority in setting transition objectives. Finally, the report calls for the planning process prior to the transition to be inclusive and transparent.[47]

Another proposal to help societies weather the transition was prepared by London School of Economics researchers Fergus Green and Ajay Gambhir. It focuses on a suite of policy alternatives that ensure that "low-carbon transitions be implemented justly, equitably, and politically smoothly."[48] The obvious policy measures involve direct grants or other forms of cash payments / low-interest loans to support displaced workers during their professional transition. In some cases, early retirement may be the most feasible alternative if retraining programs hold little promise for an individual at the end of their career.

The British researchers emphasize an important point that is often overlooked: the transition to a low-carbon economy will probably cause damages which are not only financial in nature. At the community level, this can include loss of social support networks and basic institutional structures through which communities operate. At the individual level, beyond salary loss, people will suffer diminished self-esteem and miss a connection with the people, institutions, or places with whom they have long been attached. Accordingly, when compensation packages are devised, they should contain support beyond narrow financial losses. For instance, for communities disproportionately harmed by the transition (e.g., coal-mining communities), in-kind social, cultural, or civic environmental/amenity benefits may help ameliorate the associated trauma or disillusionment. The British report identifies four types of disruption responses that need to be part of general climate policy:

- *Compensation*: Beyond existing unemployment payments, safety nets can be expanded to include unconditional financial payments that offer compensation, including early retirement packages, to offset financial losses incurred by displaced individuals. At the community level, social services can be subsidized.
- *Exemptions*: This generally is relevant at the firm level, in order to support companies and employers in situations when full enforcement of low-carbon standards would completely undermine the viability of businesses.
- *Structural adjustments*: These include a broad variety of wage subsidies to ease the shift into new fields of work. Government support can cover employment search costs, training programs, or relocation. At the community level, programs can include subsidized loans to overhaul factories that are no longer profitable or special grants to affected industries for research and development.
- *Comprehensive adaptive support*: Financial and nonfinancial measures must include the full range of adaptive support associated with structural adjustment. For individuals, in-kind assistance falls in this category, including continuing education, psychological counseling, career counseling, and transportation services. Affected communities, for example, can receive new cultural centers or infrastructures to compensate for their shrinking tax base.[49]

Such climate-transition strategies can be seen as natural extensions of environmental justice programs. In 1994, a US executive order recognized that all communities should enjoy the same degree of protection from environmental and health hazards, and equal access to the decision-making process. According to the federal policy: "Environmental Justice seeks to prevent the disproportionate burden of environmental and health hazards on marginalized and vulnerable communities, including low-income communities and communities of color."[50]

Climate disruption is fundamentally different. Here, populations suffer disproportionate financial loss due to the *abatement* of environmental hazards. Yet, in pursuing environmental objectives, policy makers always need to be aware of social ramifications. Clear philosophical guidelines can help decision-makers prioritize limited resources to confront climate-disruption challenges.

The *maximin principle* is a well-established criterion for allocation of societal resources proposed by Harvard philosopher John Rawls. Rawls posited that a just society should be designed to maximize the well-being of those who are the worst off.[51] In other words, in allocating resources, social and economic inequalities should be corrected by prioritizing benefits for the most disadvantaged in society. In designing a strategy to address disruption from decarbonization, this principle offers a strong conceptual anchor for government interventions.

Already, many social policies embrace this perspective. US president Joe Biden established an Interagency Working Group to make sure that government resources are made available to revitalize the economies of coal, oil, gas, and power plant communities. As a result, the US Economic Development Administration Coal Communities Commitment set aside $300 million to support coal-mining communities in Alaska, New York, Oklahoma, Pennsylvania, and West Virginia. Funding is available for activities and programs that support economic diversification, job creation, capital investment, workforce development, and re-employment opportunities.[52]

America's 2022 flagship climate-mitigation legislation, the Inflation Reduction Act, does not leave it to other social programs to sort out "any mess" that its ambitious agenda might create. The Act defines "energy communities" that include brownfield sites (lands abandoned due to pollution from industrial use) and metropolitan areas with meaningful employment or 25 percent greater tax revenues from extraction, processing, transport, or storage of coal, oil, or natural gas. It also includes locations where a coal mine has closed or a coal-fired plant has been retired. Four billion dollars is set aside for projects in these communities; $1.6 billion is reserved for projects in designated coal communities. The Inflation Reduction Act also sponsors investments in energy projects and infrastructure through the Department of Energy's Tribal Energy Loan Guarantee Program. It even provides special rebates for efficient appliances and clean energy on tribal lands.[53]

This hardly guarantees that adversely affected communities will end up with the compensation that the US Congress intended. Support often takes the form of tax credits that may not be relevant for disadvantaged households or municipalities. And to receive these, applicants must navigate a bureaucratic maze. IRA procedures require local taxpayers to first submit concept papers to the Internal Revenue Service. After these documents are reviewed, the Department of Energy will respond with a letter encouraging or discouraging a full application, which must be received within forty-five days. Applications must demonstrate threshold commercial viability and emissions reductions along with strengthening US supply chains and domestic manufacturing.

Only then will a series of social benchmarks be applied in order to select projects for funding. These include providing the greatest domestic job creation; providing the greatest net impact in avoiding or reducing air pollutants or anthropogenic emissions of greenhouse gases; having the greatest potential for technological innovation and commercial deployment; having the lowest levelized cost of generated or stored energy; and having the shortest project time from certification to completion. If a community passes this "audition," well, that's just the start of the paperwork.[54] Clearly, the government's intentions are benign. Every stage and each criterion makes perfect sense. But the cumulative headache may simply become too much.

Compassion for needy individuals is important. But when considering economic turbulence, policy makers ultimately need to maintain a dispassionate perspective. Labor markets in the developed world face a tsunami. But it has *little* to do with the decarbonization and *everything* to do with the occupational upheaval that artificial intelligence and automation bring with them. A recent Goldman Sachs study predicts that 300 million jobs will be lost worldwide or degraded by artificial intelligence. For many companies this is a good thing—allowing them to cut costs. As a result of AI, global GDP stands to grow by as much as 7 percent.[55]

AI disruption is hardly a "futuristic scenario." The phenomenon is well underway. In 2021, the National Bureau of Economic Research issued a report which attributed 50–70 percent of the decline in wages among blue-collar workers over the past forty years to automation. Factory and fast-food restaurant workers; taxicab and truck drivers; telephone receptionists and even some doctors are being replaced by robots, which never grow tired, never get bored, are never distracted, and never strike for higher wages.[56] Slowly but surely, many forms of human labor are becoming dispensable. The loss of jobs caused by renewable energy or a low-carbon food system is trivial by comparison.

There is little disagreement that the emerging, low-carbon economy will benefit far more people than it will leave behind. Yet there is no room for indifference when considering the many people and communities who will find that the "pursuit of a greater good" has been anything *but* good for them! It is important that a "just transition" find expression in climate policy, even if it does not fully reach the lofty aspirations of a Scandinavian think tank. Society should strive to minimize structural injustices and implement well-designed, preemptive interventions to ease the collateral damage caused by climate tech disruption. But, ready or not . . . here it comes.

CHAPTER 10

Development and Decarbonization

It was November 2016 and world leaders were convening, yet again, to discuss the state of the Earth's climate, this time in Marrakesh, Morocco. By chance, Luc Gnacadja bumped into Patrice Talon, the president of Benin, a political colleague from his West African home. Trained as an architect, Gnacadja had been appointed Benin's minister of environment in 1999. Eight years later, he was elected executive secretary of the United Nations Convention to Combat Desertification, catapulting him at a relatively young age into a leadership role in the international environmental community. During six years in the position, Gnacadja forged new connections between the UN conventions on desertification, biodiversity, and climate, emerging as one of the most eloquent and thoughtful leaders on the front lines of global efforts to pursue a sustainable future.[1]

Three years after leaving the post, Gnacadja remained active on environmental issues. During this unexpected encounter at COP22, he gave his former boss an earful about Benin's half-hearted commitment to renewable energy. At the time, Benin imported as much as half of its electricity from oil-rich Nigeria.[2] Yet its development strategy only planned for 10 percent of the country's electricity to come from solar sources. Gnacadja felt that Benin could be far more energy-independent and environmentally conscientious. "We should set higher renewable goals," he urged the president. "We should be investing to catch up with what's happening around the world. It's a good investment in the future."

President Talon thought for a second and replied, "You know, energy is the blood of the economy. Without it we simply can't develop. But solar energy is just not reliable. When you consider all the resources that need to go into buying solar systems, the premiums you have to pay on the loans that we would need to take—at the end of the day, the investment is just too costly."

Gnacadja was not easily dissuaded, and the president ultimately agreed to bring in an international expert who worked with local electricity providers to see how to better integrate a mix of wind and solar into the grid. The process culminated in a high-level workshop, which led to approval of a new sustainable energy strategy. Today, Benin is closing in on its new 30 percent renewable target.

Many countries in Africa, and in the Global South more broadly, face similar challenges as those of Benin. Gnacadja describes the dynamics that have often left low-income countries out of the renewable revolution: most low-carbon energy systems require heavy investment up front, with savings only to be realized in the future. On the other hand, providing power from fossil fuels requires little initial investment, even if it is costly in the long term.

Gnacadja is a seasoned diplomat, but he is uncharacteristically forthright when pointing to the source of the problem:

> We have 60 percent of the world's potential agricultural land. Africa could feed the world. But we remain the poorest continent and need to import food. It's the same with energy. We have the sunlight that the world dreams about. But 60 percent of the people are without electricity. The barrier is financial resources. It certainly isn't technology. . . . If you ask me why we aren't moving at the pace we should, well, maybe it's because the institutions responsible for global development, like the IMF and the World Bank, that were set up in 1945, are no longer fit for the job.[3]

Ever since global warming became a confirmed phenomenon, attitudes among African political leaders toward climate mitigation have been largely consistent. There is justifiable bitterness at the core of the "climate justice" narrative: it is simply unfair that those who contributed so little to creating the problem are the most adversely affected by its impacts. Even though, today, 16 percent of the world's population lives in Africa, the World Economic Forum attributes a mere 4 percent of present global emissions to Africa's economic activities.

As future temperatures rise, things in the Global South will get worse. The World Bank projects that by 2050, "Without concrete and development action, climate change could lead more than 216 million people in six regions to migrate in their own countries. . . . Of these, 86 million would be sub-Saharan Africans. . . . Tanzania will see the highest number of internal climate migrants reaching 16.6 million, followed by Uganda with 12 million."[4]

And yet it is a terrible mistake to continue harboring the belief that the climate crisis is a problem that only rich countries need to solve. Developing countries must become partners. Here's why: In 1950, 177 million people lived in sub-Saharan Africa. By 2020, seventy years later, the number on that part of the continent had increased eightfold, reaching 1.2 billion. The United Nations projects that by 2050, the number living there will cross 2 billion, and then, the population will probably double again. That means that 4 billion people will be sharing Africa's resources by the end of the century, almost half the people on the planet.

The European Union recently published its "Energy Projections for African Countries." The prognosis for population and development is unequivocal: by 2065, if present patterns of energy production persist, African countries will release 3.4 billion tons of CO_2—25 percent more than all of Europe emits today. Moreover, Europe's greenhouse-gas emissions are rapidly declining: already in 2020, the European carbon footprint was 25 percent lower than it had been thirty years earlier. As technology evolves and net-zero strategies become operational, the decrease will be far more dramatic. The international community does not need a prophet to see the future and understand the importance of access to low-carbon technologies in the Global South. Africa needs to be part of the solution. For this to happen, there must be far greater international assistance.

From the outset of global environmental agreements, an uncomfortable tension existed between wealthy and low-income countries. Leading up to the United Nations 1972 Stockholm Convention (the first UN gathering to consider multilateral environmental cooperation), many representatives of developing nations did not want to attend at all.[5] From their perspective, international aid was ultimately a zero-sum game. Any support for environmental initiatives invariably came at the expense of development assistance—aid that, while scant compensation for the long and terrible legacy of colonialism, was desperately needed.

In the end, Indira Ghandi, then prime minister of India, announced her intention to attend the summit. Other developing countries grudgingly followed her lead. In a gesture of appreciation, the new United Nations Environmental

176 MAKING CLIMATE TECH WORK

Program's offices were opened in Kenya, making it the first UN agency to be based on the African continent. But the original mistrust about the international environmental agenda was never really resolved.[6]

No challenge illustrates this global divide more sharply than climate change. Arguments about the reality of climate change and its severity are no longer heard on the international stage today. Nor is there any real debate about which countries historically released the lion's share of greenhouse-gas emissions. Yet the consensus begins to break down over accountability for past emissions. Even less agreement exists about the ramifications of funding mitigation and adaptation efforts. During the 2022 UN climate conference, COP27 in Egypt, things came to a head: developing countries demanded "loss and damage" compensation from developed countries, while wealthier nations preferred to refocus international attention on reducing emissions.[7]

It is not clear whether there will ever be consensus about financial compensation for historic emissions; instead, expanding aid for decarbonization in the world's poorest countries offers a far more promising strategy for a unified response to the climate crisis. While that aid could be directed at countless initiatives, investing in low-carbon electricity could be the single greatest opportunity to address both climate change and poverty. By increasing people's access to energy, one might imagine that electrification would increase communities' greenhouse-gas emissions. But the World Bank calculates that investing 1.4 percent of a developing country's GDP in low-carbon technologies could actually reduce its carbon footprint 70 percent by 2050.[8]

Electrification could prevent the majority of food waste, reduce exposure to toxic smoke from cooking fires, enable children to study at night, improve the safety of women, expand irrigation options, provide ventilation, and offer myriad other improvements in quality of life. Scaling wind and solar, along with establishing microgrids, should be a paramount priority for international assistance.

In global forums, African leaders tend to prioritize adaptation efforts. This position is ethically justified. Yet meaningful compensation for climate damage has not been forthcoming—not for decades. It is time for a more pragmatic approach. The reality is, people of all nations have a shared interest in moving the planet to a decarbonized energy system. And the damage that climate change has wrought in Africa and Asia until now is trivial compared to the cacophony that can be expected if greenhouse gases push average global temperatures up by three or four degrees Celsius. Furthermore, if regions industrialize without the benefit of climate tech, the consequences will be felt over the entire planet.

The EU's carbon border adjustment mechanism (described in chapter 2) does not distinguish between imported goods from developed countries and those from the Global South. As of 2026, imports to Europe produced in developing countries will have to pay a tax if their carbon footprints are higher than those manufactured according to European standards. A recent report from the London School of Economics predicts that the tax will only affect prices moderately.[9] Nonetheless, it will act as a trade barrier in certain areas. Climate change and trade expert, Jonny Peters posits that the best way to ameliorate adverse impacts on developing countries would be for the EU to direct some of the revenues from the border tax to support developing countries in their efforts to decarbonize and meet international environmental trade criteria.[10]

In short, no nation should be left out of the global energy transition and the associated climate technologies. Yet real barriers exist. This chapter explores the idiosyncratic circumstances that make scaling low-carbon technology so vexing in emerging and developing economies, especially in Africa—and what can be done to overcome these obstacles.

Obstacles to Adopting Climate Technology and Innovation

Based solely on solar potential, Africa should be leading the world in installed renewable energy. According to the World Bank's 2023 *Global Solar Atlas*, the continent's solar irradiance is higher that Latin America's, leaving North America in a distant third place—and Europe even further behind. If even a modest fraction of the sunshine falling in the Sahara desert were converted into electricity, it would be sufficient to power the world many times over.[11]

Unfortunately, theory and practice remain worlds apart. The electricity infrastructure in most African countries is feeble, making the lack of renewable infrastructure especially distressing. With 6 gigawatts of installed solar capacity, South Africa has roughly half the solar capacity of the entire continent.[12] Notwithstanding Germany's very modest solar irradiance, its 83 million people have five times more solar capacity than 1.3 billion Africans.[13]

The low level of renewables is particularly regrettable because solar power can so readily be harnessed through off-grid systems. Clean, reliable power would allow millions of African families to make a quantum leap forward in their quality of life, immediately providing lighting (and nighttime reading), Internet access, refrigeration, and smoke-free heating and cooking, as well as global communications.

There are many objective reasons that explain Africa's low numbers, but most fall into one of five generic categories:

- Limited economic resources;
- Limited access to finance;
- Lack of infrastructure;
- Limited technical capacity; and
- Corruption, along with weak governance and regulation.

Addressing them constitutes an essential game plan if climate tech is going to scale and decarbonization truly take on global dimensions.

Poverty and Emissions

According to the World Bank, some 659 million people on the planet live in "extreme poverty," on less than $2.15 a day,[14] while the UN estimates the number is closer to 800 million.[15] Both institutions agree that the vast majority of these individuals live in Africa: 490 million, or one out of four people in sub-Saharan Africa, experience extreme poverty.[16] Many others may not be destitute but are simply very poor. It is impossible, therefore, to discuss policies to expedite a just transition or rapid decarbonization in developing countries without considering poverty. It informs all aspects of sustainability efforts in the Global South.

Conventional wisdom tells us that people with few resources have very small carbon footprints. While it is true that low incomes cannot support lifestyles that burn through fossil fuels, the relationship between poverty and carbon emissions is complicated. Deforestation is a case in point.

According to the London School of Economics, "Land use change, principally deforestation, contributes 12–20 percent of global greenhouse-gas emissions,"[17] a phenomenon that goes hand in hand with poverty. Empirical research in Malawi, for instance, confirms that forest intactness is inversely correlated with poverty in Africa: the southern part of the country has higher poverty rates and lower forest density (a proxy for intactness), while woodlands in the more prosperous north are in far better shape. By some estimates, in rural regions, forests provide 90 percent of the livelihoods that help those living in extreme poverty to survive. Once trees are removed, people will not enjoy critical ecosystem services.

The Democratic Republic of Congo is an extreme example. Its forests hold the fourth-largest terrestrial carbon reserve on the planet, and they are disappearing at an alarming rate: between 2000 and 2020, the country lost 3.6 percent of its total tree cover. In 2021 alone, an additional 3 million acres, an area the size of Connecticut, was extirpated. According to Global Forest Watch, such massive woodlands destruction releases 822 million metric tons of CO_2,[18] or roughly twice the entire carbon footprint of South Africa, the continent's most industrialized greenhouse-gas emitter.[19]

There are many reasons, including unregulated timber and mining industries, that explain why the extraordinary Congolese forest is disappearing, along with the carbon it contains. But the immediate causes are subsistence agriculture and the collection of wood for fuel. As of 2020, less than 20 percent of the people in the Congo had access to electricity.[20] Accordingly, 90 percent of the wood removed from the country's forests is deemed "informal," meaning collected by everyday people rather than clear-cut by industry.[21] With a per capita GDP under $50 a month, and 25 percent of the population defined as "food-insecure,"[22] little wonder that people avail themselves of their forests' free fuel to stay warm and to cook. Alain Engunda, a Congolese forestry expert, explains: "If the government doesn't provide alternatives, the need for charcoal is going to explode because if people don't have electricity, they cut trees."[23]

Deforestation is also a manifestation of a global system that has not yet been willing to monetize the value of carbon. Wanjira Mathai is the current managing director for Africa at the World Resources Institute, based in Nairobi, Kenya. Mathai sees deforestation as a symptom of a broader problem of climate governance:

> The situation throughout Africa and beyond is that these incredible forests are not appreciated for their critical role in sustaining life. This calls for global markets to establish a fair price for carbon. That's really the only way that these forests will be protected. Markets need to understand their biodiversity and carbon value, and not just see them as timber.[24]

Poverty is also at the heart of the food waste, which is responsible for 8–10 percent of global greenhouse-gas emissions.[25] In the United States, between 30 and 40 percent of all food—133 billion pounds, valued at $161 billion—is thrown away each year.[26] In sub-Saharan Africa, rates are surprisingly similar but for very different reasons. Without refrigeration, farmers have no way to get produce to market before it goes bad, a problem that could be addressed through expanded electrification initiatives. While the technical solution in this case is relatively simple, the issue points to a deeper pathology. Without confronting poverty itself, progress will be elusive.[27]

The problem goes beyond renewable electricity. Lack of finance means that common-sense investments simply don't happen. Drip irrigation is another example of a missed opportunity. Some 70 percent of the planet's freshwater is utilized for irrigation by agriculture.[28] The World Bank estimates that in Africa, only about 4 percent of cultivated lands are irrigated.[29] This mostly involves inefficient flood irrigation that often causes waterlogging and salinization in soil. But even inefficient irrigation can boost food production by 50 percent. Despite its enormous advantages in terms of water savings, improved yields, and reduced weeds and diseases, only about 3 percent of irrigation worldwide is based on a drip system.[30] Adoption in Africa is much lower, far below 1 percent.

This has adverse implications for climate-mitigation efforts. With food systems producing a third of global emissions,[31] reducing agriculture's methane footprint is critical. A report by the United Nations Environmental Program from 2021 estimates that 40 percent of global methane (CH_4) emissions come from agricultural activities.[32] That's where state-of-the-art drip irrigation comes in. A team of Spanish scientists measured methane emissions from rice fields in 2023. They discovered that subsurface drip irrigation systems decrease CH_4 emissions by 80 percent—and CO_2 emissions by 41 percent, relative to conventional surface irrigation.[33] In Africa, rice is the second-largest source of local calories, with per capita consumption in the Global South among the highest in the world.[34]

Drip irrigation is usually not thought of as "climate tech." Nonetheless, investing in drip irrigation systems would not only produce considerable savings to farmers and reduce water scarcity in arid regions; it would also significantly reduce methane emissions. Unfortunately, in most developing countries, farmers simply do not have the capital to finance the up-front cost of installing a relatively inexpensive "family drip" irrigation system that requires no electrical infrastructure.

What programs are effective in reducing extreme poverty? Clearly, a book about climate innovation cannot do justice to a policy problem as wicked as poverty eradication. But neither can it be ignored when discussing the challenge of scaling climate tech to Africa. There is much that can be done to make development more efficacious and sustainable so that low-income communities can be part of the low-carbon transition.

Aid will be critical, given the current state of the continent's economies, political systems, and historical legacies. Aid is not throwing "good money after bad." When implemented carefully, it works. The increase in development assistance from governments around the world of nearly 10 percent between 2000 and 2013 is credited with significant progress, especially in improving health systems in low- and middle-income countries. For instance, during this period, deaths from malaria fell by 47 percent. Beyond the intrinsic value of reduced human suffering, the economic benefits were enormous. Immunization campaigns, infrastructure upgrades, and education programs have also meaningfully contributed to poverty relief.

Another significant, though admittedly sensitive, way to improve health, relieve poverty, and mitigate climate change is to reduce fertility rates. Empirically, fertility decline in the Global South typically precedes increases in wealth, with per capita income accelerated when fertility falls to between two and three births per woman.[35] Indeed, throughout Africa, fertility rates are consistently highest in the most impoverished areas and lowest in the most prosperous regions. For instance, with a per capita annual GDP of roughly $15,000, the Seychelles is the wealthiest country in Africa. Its total fertility rate (the average number of children born to a woman during her lifetime) is only 2.2, very close to sustainable replacement levels. South Africa's $7,000 average yearly income hides dramatic earning disparities. Nonetheless, its future economic prospects appear promising, *inter alia*, because of a 2.4 birthrate. Families in Botswana have only 2.8 children on average—with per capita income of $6,800. Meanwhile, the Democratic Republic of Congo has fertility levels of 6.2, with average annual incomes of $577. Women in Chad average 6.3 births—while, on average, families earn a mere $685 a year per capita.

Innumerable successful programs have shown that well-run, voluntary family planning initiatives produce a slow but steady increase in income levels. In other words, government decisions to make contraception available, giving women autonomy over their bodies, constitute a positive step toward prosperity.[36] Such programs are also important for climate mitigation, since population growth is one of the major drivers of deforestation.

Today, for instance, the Democratic Republic of the Congo has a population of 101 million. It is expected to double by 2050.[37] If poverty and energy patterns persist, an area of forest the size of New Jersey will disappear *every year*. The population issue has been controversial for decades, bringing up issues of equity, gender, race, and power. Due to the prevailing *Zeitgeist* of the past decades, the absence of meaningful policy discussions is likely to continue for the foreseeable future. All sides should be able to agree, however, that growing populations make it harder for countries to catch up to energy and development needs. It is much easier to reduce greenhouse-gas emissions when human population is stable. All can also agree that universal access to electricity, based on renewables and microgrids, would be a good thing for both human rights and climate change. But for this to happen, people need access to credit.

Access to Finance

The financing conundrum is a major obstacle to decarbonizing rural areas. Priscilla Achakpa, executive director of the African NGO Women Environmental Programme laments: "For the communities to be able to get electricity, they have to receive loans. But then they have to pay them back. I don't know of any place where electricity is free. Minigrids are very expensive. One of the central reasons that people in developing countries are locked into a poverty cycle is that their access to capital is very, very limited. In 85 percent of the rural communities, people still have to rely on generators and fossil fuels."[38]

Beyond the dearth of capital for individuals, the credit gap for medium and small African enterprises is estimated to be a staggering $5 trillion,[39] a figure that is directly linked to investors' perception of African countries as a high risk. Caution is built into commercial banks' narrow perspective and modus operandi. The result is that innumerable, potentially profitable climate projects do not secure funding. From a global standpoint, the resulting shortfalls in financing exacerbate emissions and climate-related risk.

A study of 193 certified financial analysts explored the reasons why many investors tend to avoid Africa. The barriers to investing were many. The most commonly cited were a lack of sufficient information and illiquid markets—singled out by 29 percent of respondents; insufficient legal protection for investors, which bothered 23 percent of the analysts; corruption was named by a full 21 percent; and 15 percent were unsettled by perennial political risks. Interestingly, only 2 percent felt that excessive regulation and bureaucracy were the problem.

Investment is particularly tricky because developing countries are prone to large budget deficits that lead to dramatic depreciation of currencies. Low-carbon infrastructure generally needs to be paid off over time. By the time payments to cover infrastructure are made, they may not be worth much unless indexing is very tight. Consider the two-year period between 2020 and 2022: Zimbabwe's RTGS dollar value dropped by 97 percent; the Sudanese pound fell by 84 percent; Nigeria's naira and Ghana's cedi lost 48 percent and 44 percent of their value, respectively.[40]

To create a revenue stream for strapped communities, two small-scale approaches have proved promising: microfinance programs, which offer loans to individuals with limited credit, and the "pay-as-you-go" finance model that proved so critical in expanding mobile phone use in Africa. (The number of African cell phone holders surpasses that of Europe and the United States.[41])

Pay-as-you-go packages already allow millions of people to offer a small down payment on photovoltaic systems, followed by regular outlays that cover equipment and energy consumed. Payments can be processed through mobile money or other electronic means. No bank accounts are required. The business model relies on the money made by entrepreneurs or by companies charging interest on the cost of the equipment and energy consumed.[42]

A range of solar products, such as solar panels, lights, and TVs, can already be paid for in small installments. Using this approach, thus far, M-KOPA has sold over a million solar systems in Kenya, Tanzania, and Uganda; Lumos Global marketed over 100,000 solar systems in Nigeria; and Off-Grid Electric, over 100,000 solar systems in Tanzania. The potential market is enormous: 70 million households that lack access to electricity could be part of such a system.

But for more people to benefit from this model, the impossible economic reality faced by potential subscribers needs to be recognized. In India, microfinancing proved to be far more effective when participants were allowed to delay initial payments, even by a few months, rather than having to repay loans within a week.[43] Structuring fees to allow new electricity systems to generate financial benefits (e.g., refrigeration of vegetables) is just common sense. These companies need to be strengthened by direct grants or other financing mechanisms.

Infrastructure

Ibrahim Mayaki, the director of the African Union Development Agency, cites some very disconcerting figures:

- Only 38 percent of Africa's population has access to electricity;
- The penetration rates for the Internet is below 10 percent;
- Only a quarter of the roads are paved; and
- Poor highways, rail, and port facilities add 30 to 40 percent to the costs of goods traded across the continent.[44] To get off the ground, renewable energy projects require a reasonable road network so that trucks can carry heavy capital equipment to designated sites.

Rural areas are particularly crippled by the woefully insufficient infrastructure, which, according to the World Bank, cuts business productivity by 40 percent.[45] Long distance transmission lines for electricity are a case in point. These cables are rarely found in rural regions, where over half of Africans live, leaving most people on the continent without electricity.[46] But today, decentralized systems, such as solar (or wind) power based on minigrids can supply electricity to a localized group of prosumer customers. Rather than building expensive lines to polluting coal plants, communities can jump straight to stand-alone, renewable solutions.

That's why it is ironic and tragic that the price of solar installation is higher in Africa than in Europe, Asia, or North America. Professor David Mutekanga, an expert in ecology and environmental management at Uganda's Bugema University, expresses the commonly felt frustration at the high prices for renewable systems: "Photovoltaics are not made locally, and they turn out to be very expensive. Buying solar systems includes the costs of transporting them. And then, when they reach here, they are taxed. Most local communities are finding it simply impossible to purchase solar panels."[47]

At present, only 5 percent of Africa's electricity generation comes from solar or wind energy. Yet a recent assessment by the International Energy Agency suggests that Africa could quickly catch up and close the solar price gap.[48] Since most of the cost of solar comes from labor rather than equipment, and labor costs are low in Africa, the continent could potentially have the cheapest installed rate of solar electricity in the world.

The IEA also calculates that for the entire continent to have access to electricity by 2030, 90 million people need to get connected to an electricity source every year—three times the present rate. There are different ways to achieve this. Several studies confirm that extending the national grid is usually the most cost-effective strategy for urban areas. In rural regions, where 80 percent of energy-deprived Africans live, minigrids and stand-alone solar-electricity systems make the most sense.

Perhaps the most promising path forward for rapid expansion of green electrification is through the engagement of independent power producers (IPPs) as part of public–private partnerships. IPPs bring several advantages to such projects: development experience, capital, and technical know-how. Between 2000 and 2022, some 350 independent power projects in 36 African countries were either built or reached financial close; 85 percent of them were based on renewables.[49]

Yet getting electricity infrastructure projects up and running is never easy and can be especially exasperating in developing countries. Most of the money for the investment must be paid up front, and profits are only earned from the savings over time. In frontier markets like those in Africa, borrowers must often rely on international lenders, which has proven time and again to be prohibitively expensive.[50]

Beyond financing questions, there are difficulties associated with connecting to local utilities, which in Africa are notoriously unreliable. When the World Bank assessed the solvency of utilities in thirty-nine African countries, it found that all but two were regularly unable to recover their costs, often because their customers are unable to pay their bills.[51] As with many issues discussed in this chapter, the infrastructure problem comes back to persistent poverty. Strengthening the reliability of the grids and the institutions responsible for maintaining them should be another key priority for international aid.

Technical Capacity

Africa is littered with graveyards of old solar panels, water pumps, and other expensive equipment that well-meaning donors assumed local recipients needed—but for which there were no replacement parts, budgets, or people with the skills to fix them. Yet "capacity building" should be one of the most straightforward solutions to scaling climate technology in developing countries.

To begin with, international donors are particularly fond of this type of aid. It offers them the opportunity to funnel funds to their own local experts and institutions to conduct training programs under the guise of humanitarian aid. It allows them to see where their money goes. And as aid initiatives go, it is inexpensive. At the 2022 UN climate conference, forty-five world leaders whose countries produce the vast majority of greenhouse-gas emissions launched the Breakthrough Agenda. A central provision in the Agenda involves bringing power-sector experts to developing countries to accelerate the transition to net-zero emissions.[52]

One example of a successful training program is located at "Barefoot College" in Rajasthan, India. For well over a decade, the college has offered a six-month program for women from off-grid communities across Africa to learn the fundamentals of solar electricity and photovoltaic installation.[53] After completing the training, the women return to their villages to electrify homes, schools, and clinics. Their communities then pay a monthly fee to ensure the maintenance and repair of the electric systems.[54] Other locally run programs offer a far cheaper alternative to flying a limited number of participants to faraway lands. For instance, the African Capacity Building Foundation has run thousands of training programs in forty-eight countries across the continent.[55] With a budget of more than $800 million, the Foundation is an important player in this very big business.[56]

The United States recently stepped up its commitment to invest in training programs as part of its climate-related aid to Africa. The Biden administration has even drafted the Peace Corps to the cause of decarbonization and capacity-building, with 700 American volunteers working in twenty-four sub-Saharan African countries in order to "increase adaptive capacities and build resilience of individuals, organizations, communities, and ecosystems, as well as reduce green-house-gas emissions and sequester carbon." Another new American initiative, Power African Jobs for Women, offers grants to NGOs in Kenya and South Africa that are confronting the many barriers to women working in the clean-energy sector.

Unfortunately, capacity-building is unlikely to come from the continent's precarious and heavily taxed private firms. A survey of stakeholders in eleven African countries found that "research and knowledge dissemination" ranked very low among priorities for private investment, though climate mitigation and affordable clean-energy projects came out somewhat higher. Despite this, the respondents cited insufficient technical skills as one of the highest barriers to sustainable development, along with "lack of motivation among the civil service" and the "fragility of the state."[57] The latter is widely recognized as a major hurdle and is the subject of the next section.

Governance, Corruption and Reliability

Historically, corruption has always been a barrier to global development, and it remains a significant obstacle to establishing new, low-carbon infrastructure. Combatting this scourge is the mission of Transparency International, a global NGO that assesses the integrity of governments around the world. Unfortunately, the countries most in need of international support tend to have the worst corruption ratings. The majority are in Africa.[58] In the oft-quoted axiom of Columbia professor and international development visionary Jeffrey Sachs: "Poverty causes corruption, and corruption causes poverty."[59]

There is obviously no magic bullet to stop corruption, a phenomenon as ancient as society itself. But along with measures such as ensuring a robust and free press, independent prosecutors and judiciaries, whistleblower protection, effective public financial management systems, and citizen participation, long-term solutions tend to depend, yet again, on increasing prosperity. International donors can play an important role, and there are many ways to ameliorate risks. One obvious place to start is the Multilateral Investment Guarantee Agency. A member of the World Bank Group, MIGA's core mission involves promoting cross-border investment in developing countries by providing guarantees (political risk insurance and credit enhancement) to investors and lenders. As most developing countries have an interest in working harmoniously with the World Bank, MIGA presumably can provide an "umbrella of deterrence against government actions that could disrupt projects and assist in the resolution of disputes between investors and governments."[60]

Finally, like everywhere in the world, the politics of energy in the Global South constitute a formidable challenge to sustainable solutions. Renewable energy advocates know well that, here too, they are not playing on an even playing field. The powerful influence of African oil interests is reflected in a study by the International Monetary Fund that found that a full 5 percent of GDP in Africa is siphoned off to provide subsidies to fossil fuel interests—including both the coal and oil industries.[61] When you add that to the prevailing political turbulence in many countries, it's little wonder that venture capitalists justifiably ask themselves: Is it really worth the risk? International aid policy needs to confront these dynamics as an important priority.

Blended Finance: A Green New Deal for Africa

In 2021, the total amount of official development assistance worldwide reached a record high of $178.9 billion. Yet this aid was only matched by $40 billion in private capital, estimated to be a mere 1 percent of the total investment needed.[62] The situation for climate mitigation is no different. In recent years, average finance for climate projects in Africa hovered around $30 billion annually, just a tenth of the amount required for African countries to meet their 2030 emissions targets.[63] Thus far, government promises and rhetoric have not been enough to catalyze investors and venture capitalists from the private sector.

It is not that there is a shortage of things to do. Due to concerns about what might go wrong in developing countries, roughly 20 percent of proposed infrastructure projects in sub-Saharan Africa, including innumerable renewable energy proposals, never get beyond the feasibility-assessment stage. And in practice, only 10 percent of proposed sustainable energy projects are actually financed to the point where construction can begin.[64]

One solution to the problem is "blended finance": capital provided by public development finance institutions (DFIs), to mobilize additional private-sector support for development projects. By transforming renewable energy projects into risk-sharing enterprises, blended financing brings together different types of capital: low-interest credit from development banks, concessional loans, and philanthropic grants, as well as commercial investments from conventional venture capital funds or impact investors.

At present, roughly a third of the funding for electricity projects in Africa invested by independent power producers comes from development finance institutions or multilateral development banks in the hope of enabling these companies to recruit private capital.[65] Many aid programs, like that overseen by the Danish Climate Investment Fund, are specifically designed to leverage private capital. Since 2014, the fund has stimulated an additional 1.3 billion Danish Krone ($117 million) for decarbonization efforts in Asia, Africa, and Latin America.[66]

It is important to emphasize that blended finance only makes up half the equation in successful green infrastructure initiatives. Host governments need to offer a stable investment environment, reasonable infrastructure (including a working and flexible electricity grid), an expeditious and uncorrupted procurement process and—ideally—a feed-in tariff to guarantee revenue to power companies.[67] Africa is no different from anywhere else: renewable portfolio standards and targets send an important message. Egypt has already announced plans to generate 42 percent of its electricity from renewables by 2035, with Morocco declaring its intentions of achieving 52 percent clean electricity by 2030.[68]

The amount of available money for "concessional capital," or financing at below-market rates, for climate related projects is slowly growing.[69] That's a good thing. But eligibility criteria for applicants also need to be more flexible. Currently, for a project to be considered for concessional lending through the World Bank and other financial institutions, the annual GDP per capita of the recipient country must be below $1,253. Yet only 12 percent of the global population reside in very poor nations, while 62 percent of the world's poor live in middle-income countries, that are home to over half of the global population.[70] There is no reason why eligibility criteria cannot be loosened.

Another strategy to reduce risks to private investors is for public finance agencies to offer loan or contract guarantees. For example, in "first-loss capital" agreements, the public partner agrees to cover delinquencies or defaults, up to a specified amount. Several financial institutions are already using this tool as part of the Africa Energy Guarantee Facility, an initiative launched by the European Investment Bank, Munich Re, and the African Trade Insurance Agency. The program guarantees support for insurance companies that cover renewable energy projects.[71] Agreements like these have been game-changing: projects are suddenly less risky for investors who are then willing to take a leap of faith and sign on.

Renewed Commitments from the Global Community

In terms of meaningful progress in mitigation, the 2009 COP in Copenhagen was largely considered lackluster. One bright spot, though, was a pledge by developed countries to mobilize $100 billion annually by 2020 for climate action in developing countries. As is often the case, it turned out to be much easier for the wealthy countries to make high-minded promises than to put their money where their mouths were.

While the international community did actually get close, raising $83 billion in 2020, three years later, donor nations were still convening to see how this commitment could be met.[72] Luc Gnacadja happened to be in Paris when French president Emmanuel Macron arranged yet another global meeting about climate financing.[73] Gnacadja does not try to hide his disappointment at the prevailing hypocrisy: "I have come to believe that the Western world is not serious when it comes to the table. They are not walking the walk. The real barrier is financial resources. When you take a close look at the money, you find that it is simply not available for African countries."[74]

Yet development abhors a vacuum. And Chinese, Turkish, Iranian, Russian, and even Indian investors are starting to fill this gap. By one estimate, China contributed $25.7 billion of Africa's $100-billion investments in infrastructure projects in 2018.[75] China's intransigence in restructuring loans has been identified as one of the major obstacles to the International Monetary Fund's assistance to some of Africa's poorest countries.[76]

It's not only the magnitude of the finances; it is the priorities. About half of the funds that do go to developing countries are directed to adaptation projects, and that proportion is increasing.[77] While adaption is very important, especially in countries where citizens lack access to clean water and sufficient food, it does little to reduce emissions. Adaptation projects will not move the continent toward the clean electrification so desperately needed if Africans are to escape the poverty trap. There is genuine concern about the ongoing sunk costs in new conventional power plants in the Global South that lock undeveloped countries into long-term dependence on fossil fuels.[78] It is little wonder that wealthy countries are more willing to fund low-carbon development.

As donor nations deliberate their climate-related aid strategy, it is well to return to the sage advice of Professor Sachs, who argued two decades ago that the key to eradicating extreme poverty was tripling international aid. In his best-selling book *The End of Poverty*, Sachs specifies four key elements critical for effective donor strategies:

- *Magnitude*—aid must be large enough to enable recipient countries to finance investment plans;
- *Timing*—aid must extend for enough time to enable recipients to draft ten-year programs;
- *Predictability*—aid must be consistent, as intermittent aid flow can jeopardize investment and economic stability; and
- *Harmonization*—aid must be integrated into global poverty-reduction strategies rather than used to support aid-agency "pet projects."[79]

Ultimately, it is critical that projects reflect meaningful collaboration with local partners. After all, it is not only Western entrepreneurs who are financing renewable energy infrastructure in Africa. There are plenty of organizations in place with whom to work. Consider the Facility for Energy Inclusion Off-Grid Energy Access Fund (FEI-OGEF), a spinoff of the African Development Bank that enjoys support from the UN Global Environment Facility, the European Commission, and the Nordic Development Facility. The Fund seeks to secure lending for a full cast of actors, including electricity distributors, manufacturers, and companies that provide credit to customers. Beyond funding, the organization offers strategic, technical, and legal support to its borrowers.[80]

It is also essential for aid to reach across the entire Global South, not only to the most prosperous countries. So far, most of the sub-Saharan African renewable energy projects initiated by international power companies have been built in South Africa, Kenya, and Nigeria. All are among the top ten economies in sub-Saharan Africa. *This means that the poorest countries on the continent, those that need most help, receive the least.*

This does not need to be the case. Jerusalem-based Gigawatt Global recently completed a 7.5-megawatt solar field in Burundi, one of the world's least-developed countries. It was Burundi's first substantial energy generation project in over three decades, making Gitega the world's only capital city powered 100 percent by solar energy during the daytime.[81] Renewable Energy for Rural Electrification in East Africa is an initiative backed by the French government that works to increase access to renewable energy in the poorest rural areas of East Africa. Its off-grid, minigrid, and clean-cooking projects in Uganda and Tanzania are funded by grants and concessional loans that leverage commercial investments. For a country like Tanzania, whose citizens had only 40 percent electricity access in 2020 (up from 5 percent in the 1990s), it could be transformational.[82]

There are no shortages of success stories to provide meaningful models moving forward. In 2015, Zambia became the first country to receive World Bank support for two photovoltaic solar electricity projects.[83] The Bank delivered financing, along with letters of credit, and then oversaw a competitive auction to attract developers and ensure transparency. The Zambian government also agreed to buy project assets or shares in the project at a specified price if its utility defaulted, which meant that the utility did not have to pony up the cash collateral to cover the letters of credit. In short, renewable electricity plants got built and the local utility gained valuable experience.[84] The approach was replicable enough to be emulated by solar projects in Senegal and Ethiopia.

For the international community to expand its financing, leadership is essential. It is heartening that the United States (perhaps due to concerns about undue Chinese influence) has gotten serious about upping its renewable game in Africa. For a decade, the USAID program Power Africa has supported access to electricity for nearly 165 million people across sub-Saharan Africa. The Biden administration then decided to expand American support for the renewable energy sector, allocating over $700 million in the coming years to create a US-Africa Clean Tech Energy Network. The initiative is designed to bring innovative US and African energy companies together and fund projects to improve supply chains, electric vehicle mobility, ecosystem conservation, food security and agriculture, energy efficiency, green hydrogen, and green finance.[85] The program has set a goal of facilitating $350 million in deals during its first five years, prioritizing public–private partnerships to electrify 10,000 health facilities in sub-Saharan Africa with renewable energy. There is every reason to believe that the initiative will succeed.

Toward Global Leadership

Despite the many challenges facing Africa, the continent has a unique opportunity to become a leader in adopting low-carbon technology, leapfrogging beyond the oil, gas, and coal paradigms that have left such a vast proportion of humanity without electricity. Kenya's progress constitutes proof of concept.

Today, renewable energy sources provide over 80 percent of Kenya's energy, increasingly making its goal of 100 percent clean energy by 2030 appear realistic.[86] It is not a coincidence that between 2009 and 2023, electrification rates in the country rose from 23 percent to 70 percent. This was largely because of the Kenya Off-Grid Solar Access Project (KOSAP) that its ministry of energy launched in 2017, providing electricity to large swaths of the country that are not served by the national grid. The project was financed by the World Bank at a relatively modest $170-million level of support.[87]

Environmental leader Wanjira Mathai points out that Africa's extraordinary solar resources have the potential to go beyond powering the continent and could be harnessed to benefit global climate efforts. In particular, she cites the momentum for creating green hydrogen from Namibia's solar and wind resources.[88] Namibia is one of the least-populated countries in Africa, with 2.5 million people on an area twice the size of California. Its enormous expanses of empty desert position it well to become the world's leading "green hydrogen center." Indeed, green hydrogen has become a central component of the country's national 2030 development strategy,[89] and in May 2023, Namibia's president, Hage Geingob, signed an estimated $10 billion deal with the German company Hyphen Energy to produce green hydrogen from Namibia's renewable sources.

To take advantage of this potential, aid programs need to prioritize decarbonization. To truly realize the incredible potential in developing countries, they will have to bring much more muscle to their work. Every country in the world signed the United Nations Framework Convention on Climate Change. Article 4 requires that countries "provide such financial resources, including for the transfer of technology, needed by the developing country Parties to meet the agreed full incremental costs of implementing measures. . . ." The meaningful impact of the World Bank's $170-million investment in Kenya should give other institutions and countries inspiration, along with concrete lessons for success.

It is time to address the acute shortfall in international finance targeting decarbonization in emerging and developing economies. If China's assistance (including its commodity-backed loans)[90] is taken out of the equation, between the 2015 Paris Agreement and 2022, the amount of climate-related aid to the Global South was stagnant. This inability to scale projects up and bring costs down in developing countries means that the capital needed for installing a solar PV plant remains two to three times higher than in advanced economies.[91] In other words, the people who can least afford to pay for renewable energy end up spending the most to get it. This doesn't make sense. Bringing climate tech to the Global South through local production centers would change this calculus.

Professor Mutekanga was recently conducting research in the field when the Ugandan grid had a power outage. For a few hours, urban centers went black. But all through the hills he could see the bright lights of villages whose residents had managed to install off-grid PV and battery facilities. "For a second, I had a vision of what might lie ahead. Those lights suggest that we really could have a brighter future."[92] More-ambitious international aid programs, along with better domestic policies, can make this vision a reality.

CHAPTER 11

Reaching Net Zero

Everywhere on the planet, signs of the climate crisis are becoming impossible to ignore: each year temperature records are broken as scorching becomes the new normal; weather patterns grow more chaotic; ice melts at unfathomable rates; fires rage out of control; cities and towns are deluged by floods of increasingly terrifying intensity and frequency.[1]

If carbon dioxide were a threshold pollutant that only caused harm when exceeding a certain concentration, like nitrates in water or sulfur dioxide in the air, then governments could simply ratchet down emissions to safe levels that both people and the natural world can tolerate. But the proverbial blanket of CO_2 enveloping the planet is not like that. If it continues to grow thicker and thicker, the Earth will just become warmer and warmer. Net-zero emissions constitute the only real strategy for stopping global warming—and then reversing it.

Carbon-neutrality requires changing pretty much everything about the way the world's economy —indeed, human civilization itself—operates. And that means that while renewable energy will be part of the solution, ultimately there is a pressing need to develop and deploy a vast array of additional innovative climate technologies. The good news is that this global technological make-over does not need to take place all at once; while civilization makes its fateful transition, the wheels of production can still run. People can still make a living. Farmers can still produce the food that nourishes humankind. The conversion will take decades. Nonetheless, the international community dare not delay dramatic steps toward decarbonization one second longer. The impacts of climate change appear to be nonlinear, which leaves humanity in a perilous twilight zone of uncertainty. The sooner the decarbonization process is complete, the sooner the Earth's systems can heal themselves.

In his recent presentations, former US vice president Al Gore never fails to conclude with a message of hope: "Scientists tell us that when we do reach a 'net-zero' equilibrium, it won't take long for things to get better. Temperatures will stop going up within three to five years. After twenty-five years, half of the of the human-caused CO_2 will fall out of the atmosphere."[2] Given its relatively short half-life, methane in the atmosphere will disappear far faster. Some of the damage wrought by climate change is irreversible, but, overall, the battle is not lost. Not yet.

Reaching a net-zero (or net-negative) equilibrium requires new technologies that are not yet available. This means that new technologies must be both developed and deployed rapidly as part of climate-mitigation policies. This book's overall message is that it can be done. But it won't happen if programs are not in place to motivate the extraordinary scientists, engineers, and managers in the climate tech ecosystem to focus on the right challenges and then help them progress as quickly as they possibly can. And this won't happen unless policies unhesitatingly put climate tech to work, replacing the unsustainable technologies that are leading the Earth into oblivion.

Electricity production has been the low-hanging fruit that most climate policies have targeted thus far. The experience from around the world is now beyond debate: it is possible to decarbonize the grid. The adoption of renewable energy in South Australia and Kenya, Germany and Denmark, California and Texas confirm that public policies can transform the economic calculus of electricity generation and clean up the world's energy mix. This is encouraging news, especially in the United States, where the Inflation Reduction Act has doubled down on subsidies for renewables. But at the same time, climate tech can only be as effective as the infrastructure supporting it. Princeton climate expert Jesse Jenkins consistently warns that if electricity grid capacity is not expanded at an unprecedented rate, half of all the projected American emission reductions—vanish.[3]

Moreover, decarbonizing electricity is only a quarter of the climate challenge. Other greenhouse-gas sources are much harder to abate. A broad range of inventions and interventions are needed to deliver the new products and processes that the world desperately needs.

In considering which policies are required to speed up the transition to net zero, there are many things that governments can do. During the past decade, most have already been tried, with varying degrees of success. By listening to the findings of evaluation research, it is possible to avoid repeating mistakes. Policy-making should have a steep learning curve.

It is already clear that climate policy is *not* an "either/or" affair. Rather, it is an all-hands-on-deck global challenge with room for many approaches. There is no silver bullet to reduce emissions and save the planet. Many tools can be found in the policy toolbox, and they all can make a contribution—if they are implemented correctly.

Monetizing carbon and creating a price that truly reflects the damage that greenhouse-gas emissions cause the planet today—and in the future—is critical to the process. This becomes increasingly important if carbon *removal* is to be incentivized. Perhaps because the public is not sufficiently aware of the dangers, politicians all over the world have been loath to set a price for carbon that reflects the true cost of emissions—a price high enough to truly induce companies and consumers to opt for low-carbon alternatives.

That needs to change. More and more citizens, especially young ones, have come to understand the consequences of inaction, and they expect carbon pricing that reflects the actual damage done to people and the planet. The new calculus may take a few years to phase in, but no source of emissions can be exempted. The atmosphere is indifferent to the political circumstances, economic exigencies, or even distributional effects behind the release of gases wafting toward the troposphere. Once they arrive, they start absorbing and re-emitting infrared radiation.[4]

In an increasingly "flat" planet, policies designed to reduce high-carbon consumption and prioritize low-carbon footprints ultimately will be far more effective than those that merely seek to rein in greenhouse-gas emissions. Without some transboundary constraints, many firms facing limits on the carbon they are allowed to release will find a way to move production elsewhere. Leakage undermines both the efficacy and the political backing for meaningful mitigation policies. Europe understood these dynamics long ago. Its bold decision—to require international competitors who want access to the EU market to pay the full carbon price that local producers are already shelling out—makes perfect sense. A great deal hinges on the successful implementation of this new carbon border adjustment mechanism and the willingness of other jurisdictions to follow suit.

When climate technologies emerge and are reasonably cost-effective, governments should not hesitate to make the desirable alternatives mandatory and ban harmful old technologies. Clearly, political prudence will usually require a phase-in period. But given the stakes, promotional campaigns and even economic incentives may be too timid and their outcomes too uncertain when proven, cost-effective solutions exist. After only a few years of enjoying uber-efficient, inexpensive LED lighting, the public today can hardly remember the protestations of a few disgruntled incandescent light zealots.

The advantages of electric vehicles are already no less self-evident: a new Chinese car, the Seagull, was unveiled in June 2023 and will cost a mere 78,000 yuan (less than $11,000). It comes in two available models: one with a range of 305 kilometers (190 miles), the other with a range of 405 kilometers (252 miles).[5] Many jurisdictions have begun to set dates for banning the sales of new gasoline- and diesel-powered vehicles. Such technology-forcing policies need to be ubiquitous. It is time to prioritize charging infrastructures and move up the date for phase-out—everywhere.

The world needs to accelerate similar transitions in the food system; in steel and cement production; in aviation and shipping. By adopting procurement strategies that prioritize goods and services with sustainable carbon footprints, governments and corporations must increasingly use their enormous purchasing power to drive the market, catapulting sustainable products toward economies of scale and lower prices that make them accessible to all.

Getting to net zero also means that people will need to change the way they live. Nudging them toward more-sustainable usage of electricity or a low-carbon diet through defaults is not only legitimate; using choice architecture is essential for helping the public adopt new, low-carbon behavior patterns.

It is also well to remember that the climate crisis is global in its dimensions. It is fine to feel virtuous if one is fortunate to live in a country that is already transitioning to a cleaner and greener way of living. But it is not a solution. The Global South wants to be part of this transformation, and it needs assistance. Prioritizing international aid for electrification with solar or wind systems is no longer simply a humanitarian gesture. It is a rational act of self-interest, and an urgent act of survival for humanity.

This is a book about bolstering climate tech. There will surely be those who question whether it makes sense to make technological innovation the sole strategy for extricating humanity and the planet from the climate crisis. That's a reasonable critique. And surely in a perfect world—and one with about a tenth of the present human population—it might be possible to merely double down on sustainable lifestyles. But, with so many billions of people justifiably striving to improve their standard of living, there is absolutely no way to stabilize atmospheric concentrations of greenhouse gases without a complete technological makeover.

This does not mean that mitigation is simply a scientific or engineering challenge. People will need to adopt low-carbon alternatives. Emissions from livestock production are expected to increase by 90 percent as diets shift to incorporate meat and the global population approaches 10 billion by midcentury.[6] Experience suggests that policies are not going to change people's eating preferences. Eliminating methane emissions from farm animals may be possible, however, if there are cultured meat or dairy alternatives that outperform cows with cheaper, healthier foods—indistinguishable in their taste and texture from the feed-lot alternative. People are also not going to stop using plastics. But public policies can ensure that the plastics of the future are biodegradable and that they sequester carbon.

Ultimately, societies need governments so that, in times of crisis, they can implement policies that deliver optimal outcomes for the common good. Human ingenuity is exceptional. When governments harness it to make vaccinations or deadly weapons, they invariably succeed. It is time for decision-makers to exploit the world's unprecedented scientific sophistication and creativity to prioritize the decarbonization of the global economy. The climate crisis will not be solved without interventions that hasten new technologies that allow people to enjoy the high quality of life to which they have become accustomed, *without* undermining the planetary systems required to support it.

Such dramatic change will not be easy. Everyone's life will probably be disrupted in one way or another along the way. Smart public policies are not only capable of making this transition happen faster—they can make it easier. Alternatively, it is of course possible to prefer a path of least resistance and blindly hope that the market will somehow deliver the technologies that are required for a carbon-neutral and inhabitable future world. But it seems like a much better idea to "choose life." By investing in innovation and making climate tech work today, policies implemented by this generation can ensure that generations to come will enjoy both a healthy economy and a healthy planet.

ACKNOWLEDGMENTS

When I came to realize in January 2023 that I was actually going to teach a course at the Stanford Graduate School of Business about public policy and climate innovation, it did not take long for me to discover that a whole lot was happening in the field. It seemed like the best way to catch up and prepare for a new class was to start researching and writing a book. That's why it makes sense to begin by thanking the inimitable Professor Bill Barnette, a world-class scholar and organizational theorist, who as of late has brought his considerable knowledge and insights to the challenges of sustainability. Bill is also the kind of person who gets things done and constitutes a *causa sine qua non* for bringing me back to Stanford, getting my course in the catalogue on time, and being a wonderful colleague throughout my most recent stays at the university.

Professor Jesper B. Sørensen, the senior associate dean for academic affairs at the business school, made me feel welcome even before I arrived, and I am grateful for his hospitality as well. More recently, Professor Larry Diamond and Dr. Kate Tyminska at Stanford's Freeman Spogli Institute for International Studies have been marvelous colleagues, introducing me to a special corner of this amazing university. Ms. Jeannine Williams remains more than just a faculty assistant: she has been a constant companion and co-conspirator. Jeannine solves every problem expeditiously, offering valuable advice about how to teach an MBA course and navigate the sundry labyrinths that one encounters in California. She knows how much I appreciate her friendship.

I cannot say enough about the students who have joined me on the journey that is listed as "GSBGEN 386." As a policy guy, I had no idea that the MBA ecosystem would be so stimulating, informative, and fun. The many discussions we have in and out of class, the tremendous case studies they prepare, and their unique visions and plans for their lives after graduating—leave me optimistic about the future. I know that these extraordinary young leaders are going to make the world a better place as they make their mark in the climate tech ecosystem. Indeed, many already are. Tal Sarig remains a most diligent, thoughtful, and talented teaching assistant who keeps me on track academically, keeps me in touch with his fellow students, and keeps our classroom Zoom studio humming.

During the course of the class we have been honored to have several eminent guest lecturers speak to us directly or Zoom in. Many of the ideas they presented are endnoted in the book. In particular, it was wonderful to learn from senior EDF economist Dr. Suzi Kerr; Stanford chemistry professor Matt Kanan; Princeton University climate policy maven Jesse Jenkins; senior advisor to the US secretary of energy, Kate Gordon; Boston University climate economist Dr. Sarah Armitage; executive chair of Manifest Energy, Alexander Karsner; and former advisor on innovation to Secretary of State Kerry, Madison Freeman. It is a true honor to host Denmark's inventor extraordinaire, Henrik Stiesal, who has probably done more than any individual in history to make wind energy the sustainable alternative it is today.

While my wife and I are at Stanford, our dear friends Josh and Sylvia Starlack become our Palo Alto family. Josh and I have been buddies since we were four years old. It remains a miraculous gift to have our lives continue to be so intertwined. And how fortunate we were to have Nick and Kathryn Hadad in town once again. From the wonders of the Alpine Inn to Seek taxonomy apps, time together with these two terrific people is always enriching. And what a pleasure to return to weekly runs with Jay Hirsch, whose musings and stories about local politics and culture are so engaging that I hardly notice the steep uphills on the "Dish" trail.

Paul and Anne Ehrlich remain dear friends. During our latest stays, we have been fortunate to be their neighbors—and it is a privilege to be just a short walk away from such a wonderful source of wisdom, wit . . . and wine. They have been affiliated with Stanford for sixty-five years, so one could not get better counsel about academic life. As they have written over fifty books together, one could not receive better counsel about publishing and how we should address the Earth's many ecological challenges. And in general, they are awfully smart about how to live life.

There were many books and articles that I encountered before I started writing that were fascinating and edifying. Three informative bestsellers, published in 2021 and 2022, first got me seriously interested in the topic of climate and technology. This veritable trifecta of tomes regarding the climate crisis includes Harvey and Gillis's absorbing primer, *The Big Fix*; John Doerr's blueprint for climate action, *Speed and Scale*; and Bill Gates's compelling polemic, *How to Avoid a Climate Disaster*. Together, they offer an excellent "big picture" from which to launch this project—before I went scurrying down innumerable rabbit holes and confronted the minutiae of sundry climate-policy instruments.

A special word of thanks for climate maven Joe Kruger, presently director of research and strategy at the Georgetown University Climate Center. For several decades, Joe and Dina Kruger have not only been close friends but also a "one-stop shopping" team for answering my many questions about climate-change policy. It is still very cool to have pals among the vaunted members of the Intergovernmental Panel on Climate Change (IPCC)—winner of the 2007 Nobel Peace Prize.

No one has helped me understand the nature of the climate tech challenge better than the extraordinary Shayle Kann. Every teacher needs a teacher—and his weekly podcast *Catalyst* never disappoints; it is ever-informative, insightful, clever, practical, and hopeful. So if you have not yet stumbled onto this outstanding resource, well, lucky you. The website sports all 100-plus free episodes that have aired since November 2021 . . . so get cracking. One day I hope to meet this remarkable man, who presumably is as savvy an investor as he is an inspirational educator. Other podcasts proved valuable, hosted by experts like Dr. Melissa Lott, Ezra Klein, and Harvard professor Robert Stavins, who features congenial discussions with a pantheon of iconic environmental economists.

This book covers a lot of ground. It was inevitable that I would get a few things wrong. That's why I was so fortunate to have so many of my colleagues, or experts that I interviewed along the way, agree to take a look at early drafts to make sure that what I was writing was consistent with their experience and expertise. Specifically, reviews and comments on different parts of the manuscript were gratefully received from David Arfin, Ornit Avidar, Brian Bartholomeusz, Trevor Berill, Michael Brauer, Dan Cass, Pam Chasek, Noah Efron, Charleen Fain-Keslar, Scott Fong, Amy Frankhouser, Larry Goulder, Lior Handelsman, Mark Jacobson, Arne Jungjohann, Finn Mortensen, Nalin Kulatilaka, Johan Lilliestam, Jane O'Sullivan, Daniel Raam, Yonina Rosenthal, Lynn Schler, Bill Slott, Karina Søgaard, Mika Tal, Robyn Tal, and Mariska Verseveld. My friend, environmental engineer Teddy Fischer, offered valuable suggestions. The extensive scholarly comments and corrections suggested by my wonderful colleague, Professor Miranda Schreuers, were superb and much appreciated.

These brilliant authorities caught many mistakes. But any errors that might have been missed along the away are clearly my own.

At Island Press, my gratitude first and foremost goes to Emily Turner, a truly talented editor: wise, patient, and forthright. Emily shepherded the book through the different stages and shared the knowledge that has come from her considerable experience in environmental publishing. This book is much better thanks to her sage interventions and suggestions. Annie Byrnes also provided excellent suggestions for revising several chapters. Jaime Jennings, the Island Press associate director of publicity, and Jen Hawse, the press's partnership manager, bring unique professional experience, solid judgment, and a passion for protecting the planet to their work. Sharis Simonian was unfailingly helpful and patient as we moved to production. It has been wonderful to be supported by people who want this book to succeed as much as I do. Over the years, I have worked with eight different copyeditors who tidied up earlier tomes. All were good at their job. But Mike Fleming, who took on this manuscript, is uniquely talented—not only as a fastidious and thoughtful writer, but also as an erudite researcher who meaningfully improved the quality of each and every chapter. In short, anyone who gets to publish a book with Island Press should feel very blessed. I surely do.

I have been fortunate for almost three years now to have the benefit of an exceptional team of parliamentary advisors: Yaara Tsairi, Shimri Negbi, and Snir Schwartz. While I am in California, they continue to keep me updated about environmental and political challenges on the home front, offering their ever-judicious counsel. My thanks also go to colleagues at the public policy department at Tel Aviv University, in particular Dorit Kerret and Amos Zehavi, for being so helpful during my time away.

Finally, as always, I am grateful to my family. As Californian ex-pats, my parents couldn't have been more supportive of this adventure. My mother, Dr. Yonina Rosenthal, found time to go over the book chapters and bring the ever-observant and analytical proficiency of a retired English professor to early drafts. Over the years, her rare editorial talent has been something on which my manuscripts have come to rely. Once again, her innumerable astute and no-nonsense comments improved the manuscript immeasurably. I cannot thank her enough. Weekly chess games with my father, David Tal, via the Internet were a pretty good substitute for our usual matches back home. How grateful we are for your continued good health and spirits! Our children, Mika and Thomas; Hadas and Yoni; Zoe and Itai were ever encouraging. Getting together intermittently during the time I was writing provided the oxygen to help me keep running what turned out to be a six-month long sprint.

My wife, Robyn, had to hear more about climate tech policy this year than she ever bargained for when she agreed yet again to set off on a sabbatical. This was our first time away as empty nesters—and how fortunate I am to have had this time together with her after a demanding political adventure. Her enduring tolerance and patience with me, especially when I get swept up in a book project, remain one of the great mysteries of our marriage. But it probably makes me the luckiest person in the world.

My daughter, Mika Tal, now works at Arcadia developing analytics tools for sustainable energy usage and saving. She has already forgotten more about climate tech that I shall ever know. I remember well a comment she threw out about a year ago during a morning run: "It seems to me that when you only deal with the problem of climate change, you can't help but get depressed. But you know, when you work in climate tech, you are developing solutions. It leaves you much more optimistic. You should try it."

She was right.

Preface
1. Zachary Liscow and Quentin Karpilow, "Innovation Snowballing and Climate Law," *Washington University Law Review 95 (2017)*: 387.

Chapter 1. Tech Policy and the Climate Crisis
1. David Smith, "Joe Biden hails Senate deal as 'most significant' US climate legislation ever," *The Guardian*, July 28, 2022, https://www.theguardian.com/us-news/2022/jul/28/joe-biden-climate-deal-senate.
2. Al Gore (@algore), Twitter post, July 28, 2022, 9:59 p.m., https://twitter.com/algore/status/1552473730458296321?lang=en.
3. Barack Obama (@BarackObama), Twitter post, July 28, 2022, 10:30 a.m., https://twitter.com/BarackObama/status/1552662824329043969.
4. Daren Bakst, "'Inflation Reduction Act' Is a Euphemism for Big Government, Socialism, and Higher Prices," *Commentary*, Heritage Foundation, August 2, 2022, https://www.heritage.org/budget-and-spending/commentary/inflation-reduction-act-euphemism-big-government-socialism-higher.
5. Tony Romm, "Inflation Reduction Act foes race to repeal climate, drug pricing programs," *Washington Post*, June 21, 2023, https://www.washingtonpost.com/business/2023/06/18/foes-inflation-reduction-act-race-repeal-climate-drug-pricing-programs/.
6. Mike Kelly, "Kelly opposes Democrats' wasteful $740 billion so-called 'Inflation Reduction Act'" (media release), August 12, 2022, https://kelly.house.gov/media/press-releases/kelly-opposes-democrats-wasteful-740-billion-so-called-inflation-reduction-act.
7. Justin Worland, "Why the World Is Protesting America's Climate Plan," *Time*, January 15, 2023, https://time.com/6247230/inflation-reduction-act-global-response-climate-trade-protectionsim.
8. Jack Loughran, "Climate policy ambitions ramping up in clean energy arms race," *Engineering and Technology*, January 26, 2023, https://eandt.theiet.org/content/articles/2023/01/climate-policy-ambitions-ramp-up-in-2022-over-clean-energy-arms-race/.
9. David Helvarg, "Our too little, too late climate action means triage more than prevention," *Los Angeles Times*, September 6, 2022, https://www.latimes.com/opinion/story/2022-09-06/climate-change-inflation-reduction-act.
10. Ronnie Greene, "Recurring red flags failed to slow Obama administration's race to help Solyndra," *Center for Public Integrity*, September 13, 2011, https://publicintegrity.org/politics/recurring-red-flags-failed-to-slow-obama-administrations-race-to-help-solyndra/.
11. Ezra Klein, "Transcript: Ezra Klein Interviews Robinson Meyer," *The Ezra Klein Show*, July 7, 2023, https://www.nytimes.com/2023/07/07/podcasts/ezra-klein-podcast-transcript-robinson-meyer.html.

12. Katie Fehrenbacher, 205, "Why the Solyndra mistake is still important to remember," *Forbes*, August 27, 2015, https://fortune.com/2015/08/27/remember-solyndra-mistake/.
13. Matthew Cantor, "Rubbish! San Francisco's $20,000 designer trash can struggles to contain trash," *The Guardian*, August 23, 2022, https://www.theguardian.com/us-news/2022/aug/22/san-francisco-cost-trash-can-struggle.
14. Goldwater Institute, "5 Examples of Government Waste That Highlight the Need for Accountability," https://www.goldwaterinstitute.org/5-outrageous-examples-of-government-waste-that-highlight-the-need-for-accountability/.
15. Bert Chapman, "Waste and duplication in NASA programs: The need to enhance U.S. space program efficiency," *Space Policy* 31 (2015): 13–20.
16. US Government General Accountability Office, "F-35 Joint Strike Fighter: Cost Growth and Schedule Delays Continue," 2022, https://www.gao.gov/products/gao-22-105943.
17. Alexander Hamilton, "Final Version of the Report on the Subject of Manufactures," 1791, https://founders.archives.gov/documents/Hamilton/01-10-02-0001-0007
18. Robert Angevine, *The Railroad and the State: War, Politics, and Technology in Nineteenth Century America* (Palo Alto, CA: Stanford University Press, 2004).
19. "US Office of Science and Technology Policy," White House website, https://www.whitehouse.gov/ostp/, accessed January 17, 2023.
20. *Japan*: Fumio Suzuki Kodama, "How Japanese Companies Have Used Scientific Advances to Restructure Their Businesses: The Receiver-Active National System of Innovation, *World Development* 35, no. 6 (June 2007): 976–90; *Israel*: Erez Maggor, "The Politics of Innovation Policy: Building Israel's 'Neo-developmental' State," *Politics & Society* 49, no. 4 (2021): 451–87; *South Korea*: Dayton Leigh, "How South Korea made itself a global innovation leader," *Nature* 581 (2020): S54–S56; *European Union*: Mattia Casula, "Implementing the transformative innovation policy in the European Union: How does transformative change occur in Member States?" *European Planning Studies* 11 (2022): 2178–204; *Germany*: David Soskice, "German technology policy, innovation, and national institutional frameworks," *Industry and Innovation* 4, no. 1 (1997): 75–96; *Ireland*: Sean O'Riain, "From Development Network State to Market Managerialism in Ireland," in *State of Innovation: The U.S. Government's Role in Technology Development*, ed. Fred Block and Matthew Keller (Boulder, CO: Paradigm, 2011), 196–216; *Denmark*: Rowan Conway, "From Competition State to Green Entrepreneurial State: New Challenges for Denmark," *Samfundsøkonomen*, 2022, https://www.djoef-forlag.dk/openaccess/samf/samfdocs/2022/2022_2/Samf_2_2_2022.pdf; *Sweden*: T. S. Adebayo et al., "Asymmetric nexus between technological innovation and environmental degradation in Sweden: An aggregated and disaggregated analysis," *Environmental Science Pollution Research* 29 (2022): 36547–64.
21. Adam Smith spoke of three factors of production (land, labor, and capital) in his classic 1776 book, *The Wealth of Nations*. During the twentieth century, entrepreneurship was added by a litany of scholars as a fourth. See: Adam Smith, *The Wealth of Nations* (New York: Shine Classics, 2014).
22. Thomas Alsop, "Nokia net sales worldwide from 1999 to 2021, *Statista*, 2022, https://www.statista.com/statistics/267819/nokias-net-sales-since-1999/.
23. Dave Lee, "Nokia: The rise and fall of a mobile giant," *BBC*, 2013, https://www.bbc.com/news/technology-23947212.

24. J. Bessant et al., "Managing innovation beyond the steady state," *Technovation* 25, no. 12 (2005): 1366–76.

25. Sanjaya Lall, *Learning from the Asian Tigers: Studies in Technology and Industrial Policy* (New York: Palgrave, 2002).

26. Lloyd Austin, "The Department of Defense Releases the President's Fiscal Year 2023 Defense Budget" (press release), US Department of Defense, 2023, https://www.defense.gov/News/Releases/Release/Article/2980014/the-department-of-defense-releases-the-presidents-fiscal-year-2023-defense-budg/.

27. Allan Bentley and Joanna Lewis, "Green Industrial Policy and the Global Transformation of Climate Politics," *Global Environmental Politics* 21, no. 4 (2021): 1–19.

28. Thomas Caplan, "The Orphan Drug Act Revisited," *Journal of the American Medical Association* 321, no. 9 (2019): 833–34.

29. National Organization for Rare Disorders, *Orphan Drugs in the United States: An Examination of Patents and Orphan Drug Exclusivity* (report), NORD, 2021, https://rarediseases.org/wp-content/uploads/2022/10/NORD-Avalere-Report-2021_FNL-1.pdf.

30. UK Ministry of Environment, Food, and Rural Affairs, *Advanced Biological Treatment of Municipal Solid Waste*, 2013, https://assets.publishing.service.gov.uk/government/uploads/system/uploads/attachment_data/file/221037/pb13887-advanced-biological-treatment-waste.pdf.

31. International Energy Agency, *Biofuels Quota Act*, 2015, https://www.iea.org/policies/5386-biofuels-quota-act.

32. European Commission, "New technique for treating wastewater through revegetation in France," 2020, https://ec.europa.eu/regional_policy/en/projects/France/new-technique-for-treating-wastewater-through-revegetation-in-france.

33. US Environmental Protection Agency, *Groundwater Technologies*, 2023, https://www.epa.gov/superfund/groundwater-technologies.

34. Nicholas Stern, *The Economics of Climate Change: The Stern Review* (Cambridge, UK: Cambridge University Press, 2007).

35. Thomas Helbling, "Back to Basics: What Are Externalities?" *Finance and Development* 47, no. 4 (December 2010), https://www.imf.org/external/pubs/ft/fandd/2010/12/basics.htm.

36. Organisation for Economic Co-operation and Development, *The Polluter Pays Principle: Definition, Analysis, Implementation* (Paris: OECD Publishing, 2008), https://doi.org/10.1787/9789264044845-en.

37. Quoted in: Paul Ehrlich, "Honoring World Population Day," *The Overpopulation Podcast*, July 22, 2022, https://www.populationbalance.org/episode-77-paul-ehrlich-transcript.

38. David Coady et al., *Global Fossil Fuel Subsidies Remain Large: An Update Based on Country-Level*, IMF Report, 2019, WP/19/89 International Monetary Fund.

39. Dominic Barton and Mark Wiseman, "Focusing Capital on the Long Term," *Harvard Business Review* (January 20, 2014), http://alliedconsult.nl/wp-content/uploads/2014/01/Focusing-Capital-on-the-Long-Term-Harvard-Business-Review.pdf.

40. Victoria Ivashina and Josh Lerner, *Patient Capital: The Challenges and Promises of Long-term Investing* (Princeton, NJ: Princeton University Press, 2019).

41. Hen Dotan et al., "Decoupled hydrogen and oxygen evolution by a two-step electrochemical–chemical cycle for efficient overall water splitting," *Nature Energy* 4 (2019): 786–95.

42. H2pro, "Our Technology," https://www.h2pro.co/technology, accessed January 18, 2023.

43. Avner Rothschild, personal interview, December 11, 2022.

44. Catherine Clifford, "After selling two software start-ups for over $1 billion, founder turns his focus to green hydrogen," CNBC website, January 25, 2022, https://www.cnbc.com /2022/01/25/h2pro-founded-by-ex-juno-viber-founder-raises-from-gates-fund.html.

45. Azeem Azhar, *The Exponential Age: How Accelerating Technology Is Transforming Business, Politics, and Society* (New York: Diversion Books, 2021).

46. Japan for Sustainability, "The Spread of Solar Power Generation in Japan," JFS Newsletter No. 70, June 2008, https://www.japanfs.org/en/news/archives/news_id027851. html.

47. Richard Swanson, *13th Workshop on Crystalline Silicon Solar Cell Materials and Processes Summary Discussion*, ed. Bhushan Sopori, National Renewable Energy Laboratory (NREL) report, 2004, https://www.nrel.gov/docs/fy04osti/35435.pdf.

48. Hal Harvey and Justin Gillis, *The Big Fix: 7 Practical Steps to Save Our Planet* (New York: Simon and Schuster, 2022).

49. Angel Adegbesan, "Solar Is Now 33% Cheaper Than Gas Power in US, Guggenheim Says," *Bloomberg*, October 2, 2022, https://www.bloomberg.com/news/articles/2022-10-03/ solar-is-now-33-cheaper-than-gas-power-in-us-guggenheim-says.

50. International Energy Agency, *Solar PV*, 2023, https://www.iea.org/energy-system/ renewables/solar-pv.

51. Sally Ho, "They've Got Beef: Beyond Meat vs. Impossible Foods Burger Showdown: What's the Difference?" *Green Queen*, September 27, 2022, https://www.greenqueen. com.hk/beyond-meat-vs-impossible-foods-burger/.

52. Ramana Nanda and Matthew Rhodes-Kropf, "Financing Entrepreneurial Experimentation," National Bureau of Economic Research, Working Paper 21278, 2015, https:// www-nber-org.stanford.idm.oclc.org/papers/w21278.

53. Benjamin Roth, teaching notes, "Prime Coalition, Catalytic Capital for Climate Innovation," Harvard Business School Case, Experimentation and Real Options in Entrepreneurial Finance HBS No. 815-056, 2022.

54. Quidnet Energy, "Our Solution, Geomechanical Pumped Storage," https://www.quid netenergy.com/solution/, accessed January 19, 2023.

55. Global Newswire, "PRIME Coalition Announces Investment in Quidnet Energy" (news release), June 16, 2015, https://www.globenewswire.com/news-release/2015/06 /16/1040520/0/en/PRIME-Coalition-Announces-Investment-in-Quidnet-Energy.html.

56. Nanda and Rhodes-Kropf, "Financing Entrepreneurial Experimentation."

57. Zachary Liscow and Quentin Karpilow, "Innovation Snowballing and Climate Law," *Washington University Law Review* 95, no. 2 (2017): 387–464.

58. Jonathan Safran Foer, *We Are the Weather: Saving the Planet Begins at Breakfast* (New York: Farrar, Straus and Giroux, 2019).

59. Robert Gifford, "The dragons of inaction: Psychological barriers that limit climate change mitigation and adaptation," *American Psychologist* 66, no. 4 (2011), 290–302.

60. Intergovernmental Panel on Climate Change, "Summary for Policymakers," in: *Global Warming of 1.5°C: An IPCC Special Report on the impacts of global warming of 1.5°C above pre-industrial levels and related global greenhouse gas emission pathways, in the context of strengthening the global response to the threat of climate change, sustainable development, and efforts to eradicate poverty*, ed. V. Masson-Delmotte et al., World Meteorological Organization, Geneva, Switzerland, 2018.

61. UNFCCC, *Outcome of the first global stocktake*, Article 28(h), 2023, https://unfccc.int/sites/default/files/resource/cma2023_L17_adv.pdf.
62. Fiona Harvey, "What are the key outcomes of Cop27 climate summit?" *The Guardian*, November 20, 2022, https://www.theguardian.com/environment/2022/nov/20/cop27-climate-summit-egypt-key-outcomes.
63. Intergovernmental Panel on Climate Change, *Contribution of Working Group I to the Sixth Assessment Report of the Intergovernmental Panel on Climate Change*, ed. V. Masson-Delmotte (Cambridge, UK: Cambridge University Press, 2021), www.ipcc.ch/report/ar6/wg1/downloads/report/IPCC_ Full_Report.pdf.
64. Martin Luther King Jr., Riverside Speech, April 4, 1967, https://inside.sfuhs.org/dept/history/US_History_reader/Chapter14/MLKriverside.htm.
65. Clayton Christensen, *The Innovator's Dilemma: The Revolutionary Book That Will Change the Way You Do Business* (New York: HarperBusiness Essentials, 2003).
66. Mark Dutz and Dirk Pilat, "Fostering Innovation for Green Growth, Learning Policy Experimentation," in *Making Innovation Policy Work: Learning from Experimentation* (Washington, DC: OECD/World Bank, 2014), 193–229.
67. J. Edler and L. Georghiou, "Public procurement and innovation: Resurrecting the demand side," *Research Policy* 36, no. 7 (2007): 949–63, https://www.ipcc.ch/site/assets/uploads/sites/2/2018/07/SR15_SPM_version_stand_alone_LR.pdf.

Chapter 2. A Global Framework for Mitigation

1. Peter Jenkins, *Mrs. Thatcher's Revolution: The Ending of the Socialist Era* (Cambridge, MA: Harvard University Press, 1988).
2. Margaret Thatcher, *The Downing Street Years* (New York: HarperCollins, 1993).
3. S. I. Rassol and S. H. Schneider, "Atmospheric Carbon Dioxide and Aerosols: Effects of Large Increases on Global Climate," *Science* 173, no. 3992 (1971): 138–41.
4. Jon Agar, "'Future Forecast—Changeable and Probably Getting Worse': The UK Government's Early Response to Anthropogenic Climate Change," *Twentieth Century British History* 26, no. 4 (2015): 602–28, https://doi.org/10.1093/tcbh/hwv008.
5. See the full text of Thatcher's UN speech in: Alon Tal, *Speaking of Earth: Environmental Speeches That Moved the World* (New Brunswick, NJ: Rutgers University Press, 119–29).
6. Intergovernmental Panel on Climate Change, *Resolutions Adopted on the Reports of the Second Committee*, IPCC, 1992, https://www.ipcc.ch/site/assets/uploads/2019/02/UNGA43-53.pdf.
7. Intergovernmental Panel on Climate Change, "History of the IPCC," IPCC, 2023, https://www.ipcc.ch/about/history/, accessed March 30, 2023.
8. United Nations, *The United Nations Framework Convention on Climate Change*, Article 4(1)(c), 1992, https://unfccc.int/resource/docs/convkp/conveng.pdf.
9. Dieter Helm, *Net Zero: How We Stop Causing Climate Change* (London: William Collins, 2020), x–xi.
10. Jonny Peters, senior policy advisor for trade and *climate*, E3G, personal communication, December 4, 2023.
11. European Union, "Application User Manual, CBAM Declarant Portal," October 2, 2023, https://taxation-customs.ec.europa.eu/system/files/2023-10/User%20Manual%20for%20CBAM%20Declarants-Release%201.1_v1.10.pdf.

12. European Commission, Taxation and Customs Union, *Carbon Border Adjustment Mechanism*, https://taxation-customs.ec.europa.eu/carbon-border-adjustment-mechanism_en, accessed December 5, 2023.
13. Meredith Fowlie, Claire Petersen, and Mar Reguant, "Border Carbon Adjustments When Carbon Intensity Varies across Producers: Evidence from California," *American Economic Association Papers and Proceedings* 111 (2021): 401–5.
14. Meredith Fowlie, personal communication, September 12, 2023.
15. California Air Resources Board, "Article 5: California Cap On Greenhouse Gas Emissions and Market-based Compliance Mechanisms," 2023, https://perma.cc/TK6X-MWJ7.
16. Meredith Fowlie, "The Whac-a-Mole Economics of Border Carbon Adjustments," *Energy Institute Blog, UC Berkeley, September 7, 2021, https://energyathaas.wordpress.com/2021/09/07/the-whac-a-mole-economics-of-border-carbon-adjustments/.*
17. Shayle Kann and Nick Wooly, "EV charging on both sides of the pond," *Catalyst*, December 1, 2023, https://www.canarymedia.com/podcasts/catalyst-with-shayle-kann/ev-charging-on-both-sides-of-the-pond.
18. Brad Templeton, "Charging Standards Can Be Easily Fixed," *Forbes*, December 19, 2019, https://www.forbes.com/sites/bradtempleton/2019/12/19/competing-electric-car-charging-standards-can-be-easily-fixed/?sh=57d049513f40.
19. Arzu Mert, Aiming Qi, Aiden Bygrave, and Henrik Stotz, "Trends of pesticide residues in foods imported to the United Kingdom from 2000 to 2020," *Food Control* 133 A (2022): 108616, https://www.sciencedirect.com/science/article/abs/pii/S0956713521007544.
20. Dieter Helm, "The Kyoto approach has failed," *Nature* 491 (2012): 663–65.
21. UNFCCC, *The Paris Agreement*, Article 2 (1)(a), 2015, https://unfccc.int/sites/default/files/english_paris_agreement.pdf.
22. Ibid., Article 4 (2).
23. David Doniger, "Paris Climate Agreement Explained: Does Congress Need to Sign Off?" Natural Resources Defense Council, December 12, 2015, https://www.nrdc.org/bio/david-doniger/paris-climate-agreement-explained-does-congress-need-sign.
24. Joeri Rogelj et al., "Paris Agreement climate proposals need a boost to keep warming well below 2°C," *Nature* 534 (2016): 631–39.
25. Ravi S. Prasad and Ridhima Sud, "The pivotal role of UNFCCC in the international climate policy landscape: A developing country perspective," *Global Affairs* 7, no. 1 (2021): 67–78.
26. Angus Naylor and James Ford, "Vulnerability and loss and damage following the COP27 of the UN Framework Convention on Climate Change," *Regional Environmental Change* 23, no. 38 (2023), https://doi.org/10.1007/s10113-023-02033-2.
27. UNFCCC, *Copenhagen Accord*, Article 8, 2009, https://unfccc.int/resource/docs/2009/cop15/eng/l07.pdf.
28. OECD, "Aggregate trends of Climate Finance Provided and Mobilised by Developed Countries in 2013–2020," 2022, https://www.oecd.org/climate-change/finance-usd-100-billion-goal.
29. A. Hattle and J. Nordbo, "That's Not New Money: Assessing how much public climate finance has been 'new and additional' to support for development," CARE Denmark, 2022, https://www.care-international.org/sites/default/files/2022-06/That%27s%20Not%20New%20Money_FULL_16.6.22.pdf.

30. UNFCCC, "Approved Projects," Green Climate Fund, 2023, https://www.greenclimate.fund/projects.

31. Green Climate Fund, "CF portfolio reaches USD 11.3 billion with new climate projects approved by the GCF Board," UNFCCC, 2023, https://www.greenclimate.fund/news/gcf-portfolio-reaches-usd-113-billion-new-climate-projects-approved-gcf-board.

32. UNFCCC, *COP27 Outcomes: Finance for Climate Adaptation*, 2022, https://unfccc.int/process-and-meetings/the-paris-agreement/the-glasgow-climate-pact/cop26-outcomes-finance-for-climate-adaptation.

33. Lindsay Maizland, "Global Climate Agreements: Successes and Failure," Council on Foreign Relations website, September 15, 2022, https://www.cfr.org/backgrounder/paris-global-climate-change-agreements.

34. Grantham Research Institute on Climate Change and the Environment, "The Climate Act [Denmark]," London School of Economics, 2020, https://climate-laws.org/geographies/denmark/laws/the-climate-act.

35. Y. Yang et al., "Mapping global carbon footprint in China," *Nature Communications* 11, no. 2237 (2020), https://doi.org/10.1038/s41467-020-15883-9.

36. Climate Action Tracker, "China submits updated NDC, confirming targets announced in September 2020," 2021, https://climateactiontracker.org/climate-target-update-tracker/china/2021-10-28-2/.

37. UNFCCC, 2022, "Climate Plans Remain Insufficient: More Ambitious Action Needed Now" (press release), October 26, 2022, https://unfccc.int/news/climate-plans-remain-insufficient-more-ambitious-action-needed-now.

38. UNFCCC, *COP26: Update to the NDC Synthesis Report*, November 4, 2021, https://unfccc.int/news/cop26-update-to-the-ndc-synthesis-report.

39. US Senate, *Summary: The Inflation Reduction Act of 2022*, https://www.democrats.senate.gov/imo/media/doc/inflation_reduction_act_one_page_summary.pdf.

40. European Commission, *Delivering the Green New Deal*, 2023, https://commission.europa.eu/strategy-and-policy/priorities-2019-2024/european-green-deal/delivering-european-green-deal_en.

41. Ibid.

42. State of Green, *A record year: Wind and solar supplied more than half of Denmark's electricity in 2020*, 2021, https://stateofgreen.com/en/news/a-record-year-wind-and-solar-supplied-more-than-half-of-denmarks-electricity-in-2020/.

43. Alon Tal, "Israel's Response to the Global Climate Crisis," *Israel Journal of Foreign Affairs* 15, no. 3 (2022): 409–14.

44. Naftali Bennett, personal communication to author, October 29, 2021.

45. Scott Stephenson et al., "Convergence and Divergence of UNFCCC Nationally Determined Contributions," *Annals of the American Association of Geographers* 109, no 4 (2019): 1240–61.

46. E. I. Come Zebra et al., "Assessing the Greenhouse Gas Impact of a Renewable Energy Feed-in Tariff Policy in Mozambique: Towards NDC Ambition and Recommendations to Effectively Measure, Report, and Verify Its Implementation," *Sustainability 13*, no. 5376 (2021), https://doi.org/10.3390/su13105376.

47. Sanita van Wyk, "Climate Change Law and Policy in South Africa and Mauritius: Adaptation and Mitigation Strategies in Terms of the Paris Agreement," *African Journal of International and Comparative Law* 30, no. 1 (2022): 1–24.

48. Minna Havukainen, Mirja Mikkilä, and Helena Kahiluot, "Global climate as a commons: Decision-making on climate change in least-developed countries," *Environmental Science & Policy* 136 (2022): 761–71.
49. UNFCCC, *First Nationally Determined Contribution Under the Paris Agreement*, September 2021, https://unfccc.int/sites/default/files/NDC/2022-06/South%20Africa%20updated%20first%20NDC%20September%202021.pdf.
50. Bangladesh Ministry of Environment, Forest, and Climate Change, "Nationally Determined Contributions, 2021, Bangladesh (Updated)," UNFCCC, August 26, 2021, https://unfccc.int/sites/default/files/NDC/2022-06/NDC_submission_20210826revised.pdf.
51. Net Zero Tracker, *Net Zero Stocktake, 2023*, June 2023, https://ca1-nzt.edcdn.com/Reports/Net_Zero_Stocktake_2023.pdf?v=1689326892.
52. Oxford University, *Net Zero Tracker*, https://www.bsg.ox.ac.uk/research/research-and-policy-updates/net-zero-tracker-report-finds-major-credibility-gaps-remain, accessed April 4, 2023.
53. International Energy Agency, *Tracking Power*, 2021, https://www.iea.org/topics/tracking-clean-energy-progress.
54. Neil Grant, "The Paris Agreement's ratcheting mechanism needs strengthening 4-fold to keep 1.5°C alive," *Joule* 6, no. 4 (2022): 703–8.
55. Alfredo Rivera et al., "Preliminary 2020 Global Greenhouse-Gas Emissions Estimates," Rhodium Group, 2021, https://rhg.com/research/preliminary-2020-global-greenhouse-gas-emissions-estimates/.
56. US Energy Information Administration, "Five states updated or adopted new clean energy standards in 2021," February 1, 2022, https://www.eia.gov/todayinenergy/detail.php?id=51118.
57. Hal Harvey and Justin Gillis, *The Big Fix: 7 Practical Steps to Save our Planet* (New York: Simon & Schuster, 2022), 41.
58. Margaret Osborne, "E.U. Agrees to Raise Its Renewable Energy Target," *Smithsonian Magazine*, April 3, 2023, https://www.smithsonianmag.com/smart-news/eu-agrees-to-raise-its-renewable-energy-target-180981913/.
59. The White House, "FACT SHEET: President Biden Sets 2030 Greenhouse Gas Pollution Reduction Target Aimed at Creating Good-Paying Union Jobs and Securing U.S. Leadership on Clean Energy Technologies" (press release), April 22, 2021, https://www.whitehouse.gov/briefing-room/statements-releases/2021/04/22/fact-sheet-president-biden-sets-2030-greenhouse-gas-pollution-reduction-target-aimed-at-creating-good-paying-union-jobs-and-securing-u-s-leadership-on-clean-energy-technologies/.
60. Jennifer Kagan, "Multiple Streams in Hawaii: How the Aloha State Adopted a 100% Renewable Portfolio Standard," *Review of Policy Research* 36, no. 2 (2019): 21741.
61. Nomvuyo Tena, "Kenya on Course to Achieving 100% Clean Energy by 2030," *ESI Africa*, May 12, 2022, https://www.esi-africa.com/energy-efficiency/kenya-on-course-to-achieving-100-clean-energy-by-2030/.
62. Carolyn Fischer and Richard Newell, "Environmental and technology policies for climate mitigation," *Journal of Environmental Economics and Management* 55, no. 2 (2008): 142–62.
63. Jenny Heeter, Bethany Speer, and Mark Glick, *International Best Practices for Renewable Portfolio Standard (RPS) Policies*, 9 NREL/TP-6A20-72798 (Golden, CO: National Renewable Energy Laboratory, 2019), https://www.nrel.gov/docs/fy19osti/72798.pdf.

64. Michael Greenstone and Ishan Nath, "Do renewable portfolio standards deliver cost-effective carbon abatement?" University of Chicago, Becker Friedman Institute for Economics Working Paper, 2020, https://bfi.uchicago.edu/wp-content/uploads/2020/11/BFI_WP_201962.pdf.

65. R. Wiser et al., *A Retrospective Analysis of the Benefits and Impacts of U.S. Renewable Portfolio Standards*, Lawrence Berkeley National Laboratory and National Renewable Energy Laboratory, 2016, NREL/TP-6A20-65005, http://www.nrel.gov/docs/fy16osti/65005.pdf.

66. Shan Zhou and Barry Solomon, "Do renewable portfolio standards in the United States stunt renewable electricity development beyond mandatory targets?" *Energy Policy* 140 (2020): 111377.

67. Janak Joshi, "Do renewable portfolio standards increase renewable energy capacity? Evidence from the United States," *Journal of Environmental Management* 287 (2021): 112261.

68. International Renewable Energy Agency, *Renewable Energy and Climate Pledges Five Years after the Paris Agreement*, IRENA, 2020, https://www.irena.org/publications/2020/Dec/Renewable-energy-and-climate-pledges.

69. United Nations Framework Convention on Climate Change, *Unleashing Renewable Energy's Full Potential*, 2023, https://unfccc.int/news/unleashing-renewable-energy-s-full-potential.

70. David Hart and Hoyu Chong, "Climate innovation policy from Glasgow to Pittsburgh," *Nature Energy* 7 (2022): 776–78.

71. University of Rochester, News Center, "Scientists hit key milestone in fusion energy quest" (news release), December 13, 2022, https://www.rochester.edu/newscenter/nuclear-fusion-energy-ignition-milestone-544292/.

72. European Commission, *Horizon Europe*, 2023, https://research-and-innovation.ec.europa.eu/funding/funding-opportunities/funding-programmes-and-open-calls/horizon-europe_en.

73. Jennifer Hahn, "COP27 carbon calculator aims to cut 'obscene' travel emissions from climate conference," *Dezeen*, November 2, 2022, https://www.dezeen.com/2022/11/03/cop27-carbon-calculator-aims-to-cut-obscene-travel-emissions-from-climate-conference/.

74. Breakthrough Agenda, *The Breakthrough Agenda Report*, 2022, https://iea.blob.core.windows.net/assets/49ae4839-90a9-4d88-92bc-371e2b24546a/THEBREAK-THROUGHAGENDAREPORT2022.pdf.

75. Valerie Volcovici and Ari Rabinovitch, "Israel and Jordan move forward with water-for-energy deal," *Reuters*, November 8, 2022, https://www.reuters.com/business/cop/israel-jordan-move-forward-with-water-for-energy-deal-2022-11-08/.

76. D. M. Hart and H. Chong, "Climate innovation policy from Glasgow to Pittsburgh," *Nature Energy* 7 (2022): 776–78, https://doi.org/10.1038/s41560-022-01113-7.

77. Paul Bonar and David Turk, "Announcing Mission Innovation," The White House, President Barack Obama, November 29, 2015, https://obamawhitehouse.archives.gov/blog/2015/11/29/announcing-mission-innovation.

78. Hart and Chong, 776–78.

79. Mission Innovation, *Missions*, 2023, http://mission-innovation.net/missions/, accessed April 3, 2023.

80. World Economic Forum, *First Movers Coalition*, 2021, https://www.weforum.org/first-movers-coalition.
81. UN Climate Change Conference, "COP26 World Leaders Summit—Statement on the Breakthrough Agenda," https://ukcop26.org/cop26-world-leaders-summit-statement-on-the-breakthrough-agenda/.
82. UNFCCC, *The Breakthrough Agenda*, 2022, https://climatechampions.unfccc.int/system/breakthrough-agenda/.
83. Ibid.
84. Breakthrough Agenda, *The Breakthrough Agenda Report*.
85. Karl Mathiesen, Zack Colman, Zia Weise, and Sara Schonhardt, "COP28 ends with first-ever call to move away from fossil fuels," *Politico*, December 13, 2023, https://www.politico.com/news/2023/12/12/newest-cop28-climate-summit-text-00131257.
86. Fiona Harvey, "The UAE consensus: a deal has been signed – but what kind of deal?" *The Guardian*, December 13, 2023, https://www.theguardian.com/environment/live/2023/dec/13/cop28-live-updates-news-agreement-outcomes-draft-text-fossil-fuels.
87. UNFCCC, *First Global Stocktake, Revised Advanced Version*, Article 27, December 13, 2023, https://unfccc.int/sites/default/files/resource/cma2023_L17_adv.pdf.

Chapter 3. Jumpstarting Research and Development

1. United Nations Framework Convention on Climate Change, "Enhancing financing for the research, development, and demonstration of climate technologies," UNFCCC, 2017, https://unfccc.int/ttclear/docs/TEC_RDD%20finance_FINAL.pdf.
2. Giada Di Stefano, Alfonso Gambardella, and Gianmario Verona, "Technology push and demand-pull perspectives in innovation studies: Current findings and future research directions," *Research Policy* 41, no. 8 (2012): 1283–95.
3. Oddvar Gorset et al., "Results from Testing of Aker Solutions Advanced Amine Solvents at CO_2 Technology Centre Mongstad," *Energy Procedia* 63 (2014): 6267–80.
4. Ketan Joshi, "The Technical Hitch," in *The Climate Book*, created by Greta Thunberg (New York: Penguin Press, 2023), 261.
5. Dani Rodrik, "Green industrial policy," *Oxford Review of Economic Policy* 30, no. 3 (2014): 469–91.
6. UNFCCC, "Enhancing financing."
7. Allan Bentle et al., "Technologies and policies to decarbonize global industry: Review and assessment of mitigation drivers through 2070," *Applied Energy* 266 (2020): 114848; and "Green Industrial Policy and the Global Transformation of Climate Politics," *Global Environmental Politics* 21, no. 4 (2021): 1–19.
8. Juhern Kim, "Assessing South Korea's Transition to Net Zero," Carnegie Endowment for Peace, 2022, https://carnegieendowment.org/2022/11/22/assessing-south-korea-s-transition-to-net-zero-pub-88426.
9. Euronews, "Boris Johnson unveils £12 billion plan aimed at creating a Green Industrial Revolution," *Euronews*, November 18, 2020, https://www.euronews.com/my-europe/2020/11/18/boris-johnson-unveils-12-billion-plan-aimed-at-creating-a-green-industrial-revolution.
10. Australian Department of Energy, Climate, the Environment and Water, *Technology Investment Roadmap*, 2022, https://www.dcceew.gov.au/climate-change/publications/technology-investment-roadmap.

11. Sue Surkes, "Government approves three-billion-shekel plan to boost climate tech," *Times of Israel*, June, 27, 2022, https://www.timesofisrael.com/government-approves -three-billion-shekel-plan-to-boost-climate-technology/.

12. Ming Ding et al., "A review on China's large-scale PV integration: Progress, challenges, and recommendations," *Renewable and Sustainable Energy Reviews* 53 (2016): 639–52.

13. Isabel Haase et al., *The Use of Auctioning Revenues from the EU ETS for Climate Action* (Berlin: Ecologic Institute, 2022), https://www.ecologic.eu/sites/default/files/publica- tion/2022/EcologicInstitute-2022-UseAucRevClimate-FullReport.pdf.

14. Congressional Research Service, *Inflation Reduction Act of 2022 (IRA): Provisions Re- lated to Climate Change* (Washington, DC: CRS, 2022), https://crsreports.congress.gov/ product/pdf/R/R47262.

15. Jesse Jenkins (Princeton University), comments at Stanford Graduate School of Business class in public policy and climate innovation, May 8, 2023.

16. Interview with Tanya Das, senior associate director for energy innovation, Bipartisan Policy Center, March 2, 2023.

17. The White House, 2021, *Bipartisan Infrastructure Investment and Jobs Act, Updated Fact Sheet* (press release), August 2, 2021, https://www.whitehouse.gov/briefing-room /statements-releases/2021/08/02/updated-fact-sheet-bipartisan-infrastructure-invest- ment-and-jobs-act/.

18. International Energy Agency, *Net Zero by 2050*, IEA, 2021, https://www.iea.org/reports/ net-zero-by-2050.

19. Jeffrey Rissman et al., "Technologies and policies to decarbonize global industry: Re- view and assessment of mitigation drivers through 2070," *Applied Energy* 266 (2020): 114848.

20. Professor Emanuel Trachtenberg, director of Israel National Institute for Security Stud- ies, personal interview, February 27, 2023.

21. Steve Pentano, head of market transformation at Rewiring America, "Ramping up the pace of home electrification," interview in *Catalyst*, podcast with Shayle Kann, Oc- tober 19, 2023, https://www.canarymedia.com/podcasts/catalyst-with-shayle-kann/ ramping-up-the-pace-of-home-electrification.

22. Anders Eldrup, director, Innovation Fund, Denmark, personal interview, April 3, 2023.

23. Allan Bentley and Joanna Lewis, "Green Industrial Policy and the Global Transforma- tion of Climate Politics," *Global Environmental Politics* 21, no. 4 (2021): 1–19.

24. Scott Woolley, "Tesla Is Worse than Solyndra," *Slate*, May 20, 2013, https://slate.com/ business/2013/05/tesla-is-worse-than-solyndra-how-the-u-s-government-bungled-its- investment-in-the-car-company-and-cost-taxpayers-at-least-1-billion.html.

25. Darryl Siry, "In Role as Kingmaker, the Energy Department Stifles Innovation," *Wired*, December 1, 2009, https://www.wired.com/2009/12/doe-loans-stifle-innovation/.

26. Dan Senor and Saul Singer, *Start-up Nation: The Story of Israel's Economic Miracle* (New York: Twelve, 2011).

27. Israel Innovation Authority, "The Innovation Authority: Soaring achievements," https:// innovationisrael.org.il/en/reportchapter/innovation-authority, accessed March 13, 2023.

28. Hanan Brand, Israel Innovation Authority, personal interview, Jerusalem, December 27, 2022.

29. Jing Ge et al., "Mobilizing Key Stakeholders to Accelerate Clean Tech Innovation and Investment, Cleantech 2.0: Are We Getting It Right This Time?," Lawrence Berkeley National Laboratory, Energy Technologies Area, 2022, https://eta-publications.lbl.gov/ publications/mobilizing-key-stakeholders.

30. Peter Engelke, Margaret Jackson, and Randolph Bell, *Mapping Green Innovation Ecosystems: Evaluating the Success Factors for the World's Leading Greentech-Innovation Centers*, Atlantic Council, 2021, http://www.jstor.org/stable/resrep31089.

31. Victoria Krammen, "Sunrun Awarded $1.6M from the U.S. Department of Energy Sun-Shot Initiative to Develop Industry's First Fully Automated Project Platform" (press release), October, 22, 2013, https://investors.sunrun.com/news-events/press-releases/detail/143/sunrun-awarded-1-6m-from-the-u-s-department-of-energy.

32. Sunrun Investor Relations, "Sunrun Launches Shift: A New Home Solar Offering" (press release), April 12, 2023, https://Investors.Sunrun.Com/News-Events/Press-Releases/Detail/282/Sunrun-Launches-Shift-A-New-Home-Solar-Offering.

33. "Carbon Clean Solutions," Dealroom.com, https://app.dealroom.co/companies/carbon_clean_solutions, accessed June 15, 2023.

34. Climate KIC, "Europe's largest climate organization," https://czechia.climate-kic.org/, accessed June 15, 2023.

35. Climate KIC, "Our Start-ups," https://www.climate-kic.org/our-community/our-start-ups/, accessed June 15, 2023.

36. EIT Climate-KIC, "EIT Climate-KIC-supported Climeworks raises €27 million to commercialise CO_2 removal technology," 2018, https://www.climate-kic.org/news/eit-climate-kic-supported-climeworks-raises-e27-million-to-commercialise-co2-removal-technology/.

37. Tado, "tado° products for your heating system," 2023, https://www.tado.com/gb-en/products#3f10l9EB7fMK0cAZSD1l4I.

38. Climate KIC, "Smart home vendor tado° raises 20 million euro to fuel US expansion," May 19, 2016, https://www.climate-kic.org/news/smart-home-vendor-tado-raises-20-million-euro-fuel-us-expansion/.

39. Tomkat Center, "Center Leadership," https://tomkat.stanford.edu/, accessed June 16, 2023.

40. Brian Bartholomeusz, personal interview, April 10, 2023.

41. Ibid.

42. Ibid.

43. Brian Bartholomeusz, personal communication, July 21, 2023.

44. Sarfraz A. Mian, "US university-sponsored technology incubators: An overview of management, policies, and performance," *Technovation* 14, no. 8 (1994): 515–28.

45. International Energy Agency, *How Governments Support Clean Tech Energy Startups*, 2022, https://www.iea.org/reports/how-governments-support-clean-energy-start-ups/policy-insights.

46. Gideon Markman et al., "Entrepreneurship and university-based technology transfer," *Journal of Business Venturing* 20, no. 2 (2005): 241–63.

47. Maruschka Gonsalves and Jayne Rogerson, "Business incubators and green technology," *Urbani izziv* supplement (2019): 212–24.

48. United Nations Framework Convention on Climate Change, *Climate Technology Incubators and Accelerators* (Bonn: UNFCCC, 2018), https://unfccc.int/ttclear/misc_/StaticFiles/gnwoerk_static/incubators_index/ee343309e8854ab783e0dcae3ec2cfa6/c172d2f388234bdbbe3dd9ae60e4d7e9.pdf.

49. US Patent and Trademark Office, "General information concerning patents," 2023, https://www.uspto.gov/patents/basics/general-information-patents, accessed March 14, 2023.

50. Gary Smith and Jay Cordes, *The 9 Pitfalls of Data Science* (Oxford, UK: Oxford University Press, 2019).

51. State of Green, *Wind Energy: Driving the global market: White Papers for a Green Transition* (Copenhagen: State of Green, 2021).

52. Denmark Ministry of Higher Education and Science, *Green solutions of the future, Strategy for investments in green research, technology, and innovation*, Copenhagen, 2020. https://ufm.dk/en/publications/2020/filer/green-solutions-of-the-future.

53. Prof. Katherine Richardson, personal interview, March 15, 2023.

54. Ibid.

55. Finn Mortensen, director, State of Green, personal communication, July 27, 2023.

56. State of Green, *Wind Energy*.

57. Denmark Ministry of Higher Education and Science, *Green solutions*.

58. IRIS Group, 2018, *The Users' Experience of Innovation Fund Denmark*, Evaluation report, June 2018, https://innovationsfonden.dk/sites/default/files/2018-09/the-users-experience-of-innovation-fund-denmark.pdf.

59. Innovation Fund Denmark, "About Innovation Fund Denmark," https://innovationsfonden.dk/en/about-innovation-fund-denmark, accessed March 15, 2023.

60. John Doerr, *Speed and Scale: An Action Plan for Solving Our Climate Crisis Now* (New York: Penguin, 2021), 139.

61. Karina Søgaard, personal interview, March 14, 2023.

62. Edit Lulu Nielson, Green Power Denmark, personal interview, March 17, 2023.

63. European Union, "TRL Assessment," Horizon Europe NCP Portal (website), 2023, https://horizoneuropencpportal.eu/store/trl-assessment.

64. "Who We Are," Aalborg Portland website, https://aalborgportlandholding.com/en/who-we-are/global-presence/denmark, accessed March 16, 2023.

65. Statista, "Highest education completed among the population in Denmark in 2021," 2021, https://www-statista-com.stanford.idm.oclc.org/statistics/874956/educational-attainment-of-the-population-in-denmark/.

66. Denmark Ministry of Higher Education and Science, *Green solutions of the future*.

67. Claus Hoegh-Jensen, "Individual—Taxes on personal income," PwC Network, 2023, https://taxsummaries.pwc.com/denmark/individual/taxes-on-personal-income.

68. Amy Frankhouser, personal communication, July 22, 2023.

69. Lily Turaski et al., *Public policy imperative for CO_2-based bioplastic*, Final Project, GSN-BGEN 386, Stanford University, 2023, copy with author.

70. Aanindeeta Banerjee, "Carbon dioxide utilization via carbonate-promoted C-H carboxylation," *Nature* 531 (2016): 215–19, https://www.nature.com/articles/nature17185.

71. Aanindeeta Banerjee, personal interview, February 14, 2023.

72. Ibid.

73. Cristina Peñasco, Laura Anadón, and Elena Verdolini, "Systematic review of the outcomes and trade-offs of ten types of decarbonization policy instruments," *Nature Climate Change* 11 (2021): 257–65.

74. Ibid.

Chapter 4. Monetizing Carbon

1. Center for Climate and Energy Solutions, "Congress Climate History," C2ES website, https://www.c2es.org/content/congress-climate-history/, accessed July 28, 2023.

2. Michael Greenstone, personal interview, July 27, 2023.

3. Elijah Asdourian and David Wessel, "What is the social cost of carbon?" *Commentary*, Brookings Institute, March 14, 2023, https://www.brookings.edu/articles/what-is-the-social-cost-of-carbon/.

4. White House, *Technical Support Document: Social Cost of Carbon for Regulatory Impact Analysis Under Executive Order 12866*, Interagency Working Group on Social Cost of Carbon, United States Government, 2010, https://obamawhitehouse.archives.gov/sites/default/files/omb/inforeg/for-agencies/Social-Cost-of-Carbon-for-RIA.pdf.

5. Louis Lerner, "Climate change will ultimately cost humanity $100,000 per ton of carbon, scientists estimate," *UChicago News*, September 9, 2020.

6. David Archer, Edwin Kite, and Greg Lusk, "The ultimate cost of carbon," *Climatic Change* 162 (2020): 2069–86.

7. Asdourian and Wessel, "What Is the Social Cost of Carbon?"

8. US EPA, "Report on the Social Cost of Greenhouse Gases: Estimates Incorporating Recent Scientific Advances," 2022, https://www.epa.gov/system/files/documents/2022-11/epa_scghg_report_draft_0.pdf.

9. Government of Canada, "Pricing Carbon Pollution," 2020, https://www.canada.ca/content/dam/eccc/documents/pdf/climate-change/climate-plan/annex_pricing_carbon_pollution.pdf.

10. Michael Greenstone, personal interview, July 27, 2023.

11. Noah Kaufman et al., "A near-term to net zero alternative to the social cost of carbon for setting carbon prices," *Nature Climate Change* 10 (2020): 1010–14.

12. Robert Stavins, *Economics of the Environment: Selected Readings*, 7th ed. (Cheltenham, UK: Edward Elgar Publishing, 2019).

13. I. Bailey et al., "The fall (and rise) of carbon pricing in Australia: A political strategy analysis of the carbon pollution reduction scheme," *Environmental Politics* 21, no. 5 (2012): 691–711.

14. Anjani Datia, "Pricing Carbon: The Birth of British Columbia's Carbon Tax," Harvard Kennedy School Case Study, Harvard University, Cambridge, Massachusetts, 2016.

15. Ross Beaty, "Just the facts, please: The true story of how B.C.'s carbon tax is working," Smart Prosperity Institute blog, 2014, https://institute.smartprosperity.ca/content/just-facts-please-true-story-how-bc-s-carbon-tax-working.

16. Diane Toomey, "How British Columbia Gained by Putting a Price on Carbon," *Yale Environment 360*, April 30, 2015, https://e360.yale.edu/features/how_british_columbia_gained_by_putting_a_price_on_carbon.

17. Hannah Ritchie and Max Roser, "Sweden: Per capita: how much CO_2 does the average person emit?" *Our World in Data*, 2021, https://ourworldindata.org/co2/country/sweden#per-capita-how-much-co2-does-the-average-person-emit.

18. British Columbia, "British Columbia Trends in Greenhouse Gas Emissions in B.C. (1990–2018)," *Environmental Reporting BC*, 2022, https://www.env.gov.bc.ca/soe/indicators/sustainability/ghg-emissions.html.

19. Soodeh Saberian, *Essays on Environmental Economics Soodeh Saberian*, dissertation submitted to the University of Ottawa, 2018, file:///Users/alontal/Downloads/Saberian_Soodeh_2018_thesis.pdf.

20. Johan Lilliestam, Anthony Patt, and German Bersalli, "The Effect of Carbon Pricing on Technological Change on Full Energy Decarbonization: A Review of Empirical Ex-post Evidence," *Climate Change* 12, no. e681 (2021): 1–21.

21. Cristina Peñasco, Laura Anadón, and Elena Verdolini, "Systematic review of the outcomes and trade-offs of ten types of decarbonization policy instruments," *Nature Climate Change* 11 (2021): 257–65.

22. Government Offices of Sweden, "Sweden's Carbon Tax," 2023, https://www.govern ment.se/government-policy/swedens-carbon-tax/swedens-carbon-tax/.

23. Folke Bohlin, "The Swedish carbon dioxide tax: Effects on biofuel use and carbon dioxide emissions," *Biomass and Bioenergy* 15, nos. 4–5 (1998): 283–91.

24. Stanislav Shmelev and Stefan Speck, "Green fiscal reform in Sweden: Econometric assessment of the carbon and energy taxation scheme," *Renewable and Sustainable Energy Reviews* 90 C (2018): 969–81.

25. Julius Andersson, "Carbon Taxes and CO_2 Emissions: Sweden as a Case Study," *Economic Policy* 11, no. 4 (2019): 1–30.

26. European Environmental Agency, "Fuel efficiency and fuel consumption in private cars, 1990–2011," https://www.eea.europa.eu/data-and-maps/figures/fuel-efficiency-and-fuel-consumption, accessed February 10, 2022.

27. María Rodríguez, "New milestone in Norway: plug-in cars reach 93% in September," October 10, 2023, https://www.electromaps.com/en/blog/electric-car-sales-third-quarter-2023-europe.

28. María Rodríguez, "57.5 increase in BEV in Sweden", October 10, 2023, https://www.electromaps.com/en/blog/electric-car-sales-third-quarter-2023-europe.

29. Tyler Dawson, "As gas prices soar in wake of Ukraine war, Canadians could see cost of goods go up," *National Post*, March 8, 2022, https://nationalpost.com/news/canadian-economy-faces-inflationary-trend-as-gas-prices-soar.

30. Mathilde Carlier, "Daily average regular gas price in Canada throughout the war in Ukraine," 2022, *Statistica*, https://www-statista-com.stanford.idm.oclc.org/statistics/1293131/canada-daily-average-gas-prices, accessed February 8, 2023/.

31. K. Shuval et al., "Cigarette Prices and Smoking Behavior in Israel: Findings from a National Study of Adults (2002–2017*)*," *International Journal of Environmental Research and Public Health* 18 (2021): 8367, doi: 10.3390/ijerph18168367; PMID: 34444117; PMCID: PMC8394522.

32. Nalin Kulatilaka, cofounder and chief strategy officer, Nine Dot Energy, personal communication, July 4, 2023.

33. Johan Lilliestam, Helmholtz Centre Potsdam, personal interview, February 23, 2023.

34. Chris Casey, "Meat tax could help cut emissions, but must be significant to work: study," *FoodDive*, August 22, 2022, https://www.fooddive.com/news/meat-tax-beef-poultry-pork-emissions-study/630031/.

35. Food and Agriculture Organization of the United Nations, "By the numbers: GHG emissions by livestock," https://www.fao.org/news/story/en/item/197623/icode/, accessed March 3, 2023.

36. Lawrence Goulder and Andrew Schein, "Carbon Taxes versus Cap and Trade: A Critical Review," *Climate Change Economics* 4, no. 3 (2013): 1350010, https://doi-org.stanford.idm.oclc.org/10.1142/S201000781350.

37. Richard Schmalensee and Robert N. Stavins, "The SO_2 Allowance Trading System: The Ironic History of a Grand Policy Experiment," *Journal of Economic Perspectives* 27, no. 1 (2013): 103–22.

38. US EPA, *Acid Rain Program Results*, 2022, https://www.epa.gov/acidrain/acid-rain-program-results.

39. Amanda Rosen, "The Wrong Solution at the Right Time: The Failure of the Kyoto Protocol on Climate Change," *Politics and Policy* 43, no. 1 (2015): 30–58.
40. Joe Kruger and Christian Egenhofer, Confidence through Compliance in Emissions Trading Markets," *Sustainable Development Law and Policy Journal* 6, no. 2 (2006): 8–23.
41. Easwaran Narassimhan, Kelly S. Gallagher, Stefan Koester, and Julio Rivera, "Carbon pricing in practice: A review of existing emissions trading systems," *Climate Policy* 18, no. 8 (2018): 967–91, doi: 10.1080/14693062.2018.1467827; and: Simone Borghesi and Massimiliano Montini, "The Best (and Worst) of GHG Emission Trading Systems: Comparing the EU ETS with Its Followers," *Frontiers in Energy Research* 4 (2016), http://www.frontiersin.org/articles/10.3389/fenrg.2016.00027.
42. Narassimhan, Gallagher, Koester, and Rivera, "Carbon pricing in practice."
43. Regional Greenhouse Gas Initiative, RGGI website, 2023.
44. Organisation for Economic Co-operation and Development, "Carbon pricing in New Zealand Pricing Greenhouse Gas Emissions," 2022, https://www.oecd.org/tax/tax-pol icy/carbon-pricing-new-zealand.pdf.
45. J. L. Richter and L. Mundaca, "Market behavior under the New Zealand ETS," *Carbon Management* 4, no. 4 (2013): 423–38.
46. Borghesi and Montini, "The Best (and Worst) of GHG Emission Trading Systems."
47. Patrick Bayer, and Michaël Aklin, "The European Union Emissions Trading System reduced CO_2 emissions despite low prices," *Proceedings of the National Academy of Sciences* 117, no. 16 (2020): 8804–12.
48. Danny Cullenward and David Victor, *Making Climate Policy Work* (Cambridge, UK: Polity, 2021), 122.
49. Ibid., 124.
50. Damian Carrington, "EU prices crashes to record low," *The Guardian*, January 24, 2013, https://www.theguardian.com/environment/2013/jan/24/eu-carbon-price-crash -record-low.
51. European Environmental Agency, "Use of auctioning revenues generated under the EU Emissions Trading System," https://www.eea.europa.eu/ims/use-of-auctioning-reve nues-generated, accessed February 3, 2023.
52. Susanna Twidale, Kate Abnett, and Nina Chestney, "EU carbon hits 100 euros taking cost of polluting to record high," *Reuters*, February 21, 2023, https://www.reuters.com/ markets/carbon/europes-carbon-price-hits-record-high-100-euros-2023-02-21/.
53. Suchitra Mohanti, "China's National Carbon Market," Amity Research Center, Bangalore, India, 2021.
54. Lawrence Goulder and Richard Morgenstern, "Trading Carbon: China's Ambitious Initiative to Contain Climate Change," *Milken Institute Review*, April 25, 2022, https:// www.milkenreview.org/articles/trading-carbon.
55. Ibid.
56. Lawrence Goulder et al., "China's Nationwide CO_2 Emissions Trading System: A General Equilibrium Assessment, March, 2023. (Copy with author.)
57. Ibid.
58. Amy Hawkins, "China ramps up coal power despite carbon neutral pledges," *The Guardian*, April 23, 2023, https://www.theguardian.com/world/2023/apr/24/china -ramps-up-coal-power-despite-carbon-neutral-pledges.

59. David Suzuki and Ian Hanington, *Just Cool It: The Climate Crisis and What We Can Do* (Vancouver, BC: Greystone books, 2017).

60. Elizabeth Economy, "China and Climate Change: Three Things to Watch after Paris," *Forbes*, December 15, 2015, https://www.forbes.com/sites/elizabetheconomy/2015/12/15/china-and-climate-change-three-things-to-watch-after-paris/amp/.

61. Climate Action, *NER Programme: Climate Action*, European Commission, 2018, https://ec.europa.eu/clima/policies/lowcarbon/ner300/index_en.htm.

62. R. Martin, M. Muûls, and U. Wagner, "Carbon markets, carbon prices and innovation: Evidence from interviews with managers," paper presented at the Annual Meeting of the American Economic Association, San Diego, CA, 2013.

63. Á. Löfgren et al., "Why the EU ETS needs reforming: An empirical analysis of the impact on company investments," *Climate Policy* 14, no. 5 (2014): 537–58.

64. Lina Fu et al., "Do carbon emission trading scheme policies induce green technology innovation? New evidence from provincial green patents in China," *Environmental Science and Pollution Research* 30 (2023): 13342–58.

65. S. Pontogolio, "An early assessment of the influence on eco-innovation of the EU Emissions Trading Scheme," in *Environmental Efficiency, Innovation and Economic Performances*, ed. M. Mazzanti and A. Montini (London, New York: Routledge, 2010), 81–91.

66. Climate Action, "Revision for phase 4 (2021–2030)," European Commission, 2023, https://climate.ec.europa.eu/eu-action/eu-emissions-trading-system-eu-ets/revision-phase-4-2021-2030_en.

67. Danny Cullenward and David Victor, *Making Climate Policy Work* (Cambridge, UK: Polity, 2021), 2–3.

68. Allisa Rubin and Somini Sengupta, "'Yellow Vest' Protests Shake France. Here's the Lesson for Climate Change," *New York Times*, December 6, 2018, https://www.nytimes.com/2018/12/06/world/europe/france-fuel-carbon-tax.html.

69. Lucas Chancel and Thomas Piketty, "Decarbonization Requires Redistribution," in *The Climate Book*, created by Greta Thunberg (New York: Penguin Press, 2023), 405–9.

70. Simon Copland, "Anti-politics and Global Climate Inaction: The Case of the Australian Carbon Tax," *Critical Sociology* 46, nos. 4–5 (2020): 623–41.

71. Alex Frankel, "From 'axe the tax' to 'climate consensus': How Abbott reshaped our climate story," *The Guardian*, April 27, 2015, https://www.theguardian.com/commentisfree/2015/apr/28/from-axe-the-tax-to-climate-consensus-how-abbott-reshaped-our-climate-story.

72. BBC, "Australia votes to repeal carbon tax," BBC, July 17, 2014, https://www.bbc.com/news/world-asia-28339663.

73. Brookings Institute, "Event Summary: U.S. Climate Policy Toward a Sensible Center," *Brookings*, June 24, 2004, https://www.brookings.edu/opinions/event-summary-u-s-climate-policy-toward-a-sensible-center/.

74. Jens Ewald, Thomas Sterner, and Erik Sterner, "Understanding the resistance to carbon taxes: Drivers and barriers among the general public and fuel-tax protesters," *Resource and Energy Economics* 70 (2022): 101331.

75. New Zealand Ministry for the Environment, *Agriculture emissions and climate change*, 2022, https://environment.govt.nz/facts-and-science/climate-change/agriculture-emissions-climate-change/.

76. A. Tal, "Tried and True: Reducing Green House Gas Emissions in New Zealand through Conventional Environmental Legislative Modalities," *Otago University Law Journal* 12 (2009): 1–47.

77. Cameron James McLaren, "How New Zealand plans to tackle climate change: Taxing cow burps," *Washington Post*, February 1, 2023, https://www-washington-post-com.stanford.idm.oclc.org/climate-solutions/interactive/2023/new-zealand-cows-burps-methane-tax/te-solutions/interactive/2023/new-zealand-cows-burps-methane-tax/.

78. Agence France-Presse, "New Zealand Outlines Plans to Tax Livestock Gas," *VOA News*, October 12, 2022, https://www.voanews.com/a/new-zealand-outlines-plans-to-tax-livestock-gas/6787828.html.

79. Oliver Milkman, "Bill Gates backs new startup aiming to reduce emissions from cow burps," *The Guardian*, January 24, 2023, https://www.theguardian.com/us-news/2023/jan/24/bill-gates-startup-cow-burps-methane-emissions.

80. Laura Paddison, "Bill Gates backs start-up tackling cow burps and farts," *CNN*, January 24, 2023, https://www.cnn.com/2023/01/24/world/cows-methane-emissions-seaweed-bill-gates-climate-intl/index.html.

81. Rob Jackson, "Drawdown Technologies," in *The Climate Book*, 235–39.

Chapter 5. Incentives for Innovation

1. Kenneth E. Morris, *Jimmy Carter, American Moralist* (Athens, GA: University of Georgia Press, 1996).

2. Jason Bordoff, "America's Energy Policy," *Horizons: Journal of International Relations and Sustainable Development* 8 (2016): 180–205.

3. Abby Studen, *Solar Photovoltaic Technology*, Lafayette College, 2023, https://sites.lafayette.edu/egrs352-sp15-PV/policy/policy-and-presidents/, accessed June 7, 2023.

4. Catherine A. Durham, Bonnie G. Colby, and Molly Longstreth, "The Impact of State Tax Credits and Energy Prices on Adoption of Solar Energy Systems," *Land Economics* 64, no. 4 (1988): 347–55, https://doi.org/10.2307/3146307.

5. James Katz, "US energy policy: Impact of the Reagan Administration," *Energy Policy* 12, no. 2 (1984): 135–45.

6. John Wihbey, "Jimmy Carter's Solar Panels: A Lost History That Haunts Today," *Yale Climate Connections*, November 11, 2008, https://yaleclimateconnections.org/2008/11/jimmy-carters-solar-panels/.

7. David Biello, "Where Did the Carter White House's Solar Panels Go?" *Scientific American*, August 6, 2010, https://www-scientificamerican-com.stanford.idm.oclc.org/article/carter-white-house-solar-panel-array/.

8. The White House, "Inside the White House: Solar Panels," May 9, 2014, https://obamawhitehouse.archives.gov/photos-and-video/video/2014/05/09/inside-white-house-solar-panels.

9. Danyel Reiche and Mischa Bechberger, "Policy differences in the promotion of renewable energies in the EU member states," *Energy Policy* 32, no. 7 (2004): 843–49.

10. Irene Xiarchos and Brian Vick, "Solar energy use in U.S. agriculture: Overview and policy issues," US Department of Agriculture, 2011, https://naldc.nal.usda.gov/download/49148/pdf.

11. Congressional Research Service, "The Energy Credit or Energy Investment Tax Credit (ITC)," *In Focus*, CRS Reports, 2021, https://crsreports.congress.gov/product/pdf/IF/IF10479.

12. US Office of Energy Efficiency and Renewable Energy, "Federal Solar Tax Credits for Business," US Department of Energy, August 2023, https://www.energy.gov/eere/solar/federal-solar-tax-credits-businesses.

13. Robert Pitts and James Wittenbach, *Journal of Consumer Research* 8, no. 3 (1981): 335–38.

14. Paul Arnsberger et al., "History of the Tax-Exempt Sector: An SOI Perspective," *Statistics of Income*, Winter 2008.

15. Kristina Shampanier, Nina Mazar, and Dan Ariely, "Zero as a Special Price: The True Value of Free Products," *Marketing Science* 26, no. 6 (2007): 742–57.

16. David Arfin, personal interview, June 9, 2023.

17. Ibid.

18. Energy Sage, "Cost of Electricity in California," June 11, 2023, https://www.energysage.com/local-data/electricity-cost/ca/.

19. David Arfin, personal interview, June 9, 2023.

20. "What is net metering and how does it work," *Solar reviews*, February 28, 2023, https://www.solarreviews.com/blog/what-is-net-metering-and-how-does-it-work.

21. Cherrelle Eid et al., "The economic effect of electricity net-metering with solar PV: Consequences for network cost recovery, cross subsidies, and policy objectives, *Energy Policy* 75 (2014): 244–54.

22. Chelsea Schelly, Edward Louie, and Joshua Pearce, "Examining interconnection and net metering policy for distributed generation in the United States," *Renewable Energy Focus* 22/23 (2017): 10–19.

23. Beth Reid, Joe Bourg, and Devon Schmidt, "Let's Make a Deal: Non-Wires Alternatives for Traditional Transmission and Distribution?" *IEEE Power and Energy Magazine* 20, no. 2 (2022): 23–31.

24. Dan Power, "Non-Wires Alternatives Provide Benefits to Remote Communities," *Guidehouse Insights*, March, 24, 2022, https://guidehouseinsights.com/news-and-views/non-wires-alternatives-provide-benefits-to-remote-communities.

25. Ian McCarlie, "Autumn Statement 2022: Electricity generator levy a risk to renewables investment," *Out-Law*, November 17, 2022, https://www.pinsentmasons.com/out-law/news/electricity-generator-levy-a-risk-to-renewables-investment.

26. Jeff St. John, "California slashes payment to new rooftop-solar customers," *Canary Media*, December 15, 2022, https://www.canarymedia.com/articles/solar/california-slashes-payments-to-new-rooftop-solar-customers.

27. Solar United Neighbors, "Net Metering in Arizona," 2023, https://www.solarunitedneighbors.org/arizona/learn-the-issues-in-arizona/net-metering-in-arizona/, accessed June 12, 2023.

28. International Energy Agency, "Global Energy Review: CO_2 Emissions in 2021," 2022, https://www.iea.org/reports/global-energy-review-co2-emissions-in-2021-2.

29. International Energy Agency, "Snapshot of Global PV Markets 2023," 2023, https://iea-pvps.org/wp-content/uploads/2023/04/IEA_PVPS_Snapshot_2023.pdf.

30. Lior Handelsman, personal interview, December 25, 2022.

31. Internal Revenue Service, "Guidance on Provisions to Expand Reach of Clean Energy Tax Credits Through President Biden's Investing in America Agenda," US Department of the Treasury, June 14, 2023, https://home.treasury.gov/news/press-releases/jy1533.

32. "Treasury Releases Initial Information on Electric Vehicle Tax Credit under Newly Enacted Inflation Reduction Act" (press release), US Department of the Treasury, August 16, 2022, https://home.treasury.gov/news/press-releases/jy0923.

33. Rebecca Bellan, "Tracking the EV battery factory construction boom across North America," *TechCrunch*, August 16, 2023, https://techcrunch.com/2023/08/16/tracking-the-ev-battery-factory-construction-boom-across-north-america/.

34. Shayle Kann and Ethan Zindler, "Building Out a U.S. Solar Supply Chain," *Catalyst*, June 22, 2023, https://www.canarymedia.com/podcasts/catalyst-with-shayle-kann/building-out-a-u-s-solar-supply-chain.

35. Senate of the United States, Inflation Reduction Act of 2022 (Sec. 13101), §48C(e)(5)(A) and (6), 26 USC § 48C(c)(1) (prevailing wage and apprenticeship requirements), H.R. 5376, 117th Congress, 1st Session, 2022, https://www.democrats.senate.gov/imo/media/doc/inflation_reduction_act_of_2022.pdf.

36. Ibid.

37. Jesse Jenkins, *The Power Hungry Podcast*, September 15, 2020, https://robertbryce.com/episode/jesse-jenkins-assistant-professor-of-mechanical-and-aerospace-engineering-princeton-university/.

38. Kann and Zindler, "Building Out a U.S. Solar Supply Chain."

39. US Environmental Protection Agency, "Summary of Inflation Reduction Act provisions related to renewable energy," June 1, 2023, https://www.epa.gov/green-power-markets/summary-inflation-reduction-act-provisions-related-renewable-energy.

40. Internal Revenue Service, "Modified and Clarified by Notice 2023-23, Initial Guidance Establishing Qualifying Advanced Energy Project Credit Allocation Program Under Section 48C(e), APPENDIX A—Qualifying Advanced Energy Projects," 2023, https://www.irs.gov/pub/irs-drop/n-23-44.pdf.

41. The White House, *Inflation Reduction Act Guidebook*, 2023, https://www.whitehouse.gov/cleanenergy/inflation-reduction-act-guidebook/, accessed July 3, 2023.

42. John Bisline et al., "Emissions and energy impacts of the Inflation Reduction Act," *Science* 380 (2023): 1324–27.

43. The White House, "Clean Hydrogen Production Tax Credit," *Building a Clean Energy Economy: A Guidebook to the Inflation Reduction Act's Investments in Clean Energy and Climate Action, Version 2*, 2023, https://www.whitehouse.gov/wp-content/uploads/2022/12/Inflation-Reduction-Act-Guidebook.pdf.

44. Robinson Meyer, "Transcript: Ezra Klein Interviews Robinson Meyer," *The Ezra Klein Show*, July, 7, 2023, https://www.nytimes.com/2023/07/07/podcasts/ezra-klein-podcast-transcript-robinson-meyer.html.

45. "America's Government Is Spending Lavishly to Revive Manufacturing," *The Economist*, February 2, 2023, https://www.economist.com/briefing/2023/02/02/americas-government-is-spending-lavishly-to-revive-manufacturing.

46. Trevor Houser, Rhodium Group, "How Has US Industrial Policy Impacted Climatech Investment?" *Catalyst with Shayle Kann*, October 12, 2023, https://www.canarymedia.com/podcasts/catalyst-with-shayle-kann/how-has-us-industrial-policy-impacted-climatetech-investment.

47. Melissa Barbanell, "A Brief Summary of the Climate and Energy Provisions of the Inflation Reduction Act of 2022," *World Resources Institute*, October 28, 2022, https://www.wri.org/update/brief-summary-climate-and-energy-provisions-inflation-reduction-act-2022; and: John Bisline, "Emissions and energy impacts of the Inflation Reduction Act," *Science* 380 (2023): 1324–27.

48. Ibid.

49. Jennifer Runyon, "IEA: Feed-in Tariff Not a Subsidy, but Tax Credits Are," *Renewable Energy World,* November 1, 2023, https://www.renewableenergyworld.com/solar/iea-feed-in-tariff-not-a-subsidy-but-tax-credits-are/.

50. Libo Zhang et al., "The impact of feed-in tariff reduction and renewable portfolio standard on the development of distributed photovoltaic generation in China," *Energy* 232 (2021): 120933.

51. Takashi Ohba, Koichi Tanigawa, and Liudmila Liutsko, "Evacuation after a nuclear accident: Critical reviews of past nuclear accidents and proposal for future planning," *Environment International* 148 (2021): 106379.

52. World Nuclear Association, "Chernobyl Accident," 2022, https://world-nuclear.org/information-library/safety-and-security/safety-of-plants/chernobyl-accident.aspx.

53. Merethe Dotterud Leiren and Inken Reimer, "Germany: From feed-in tariffs to greater competition," in *Comparative Renewables Policy* (London: Routledge, 2020), 75–95.

54. J. P. M. Sijm, "The Performance of Feed-in Tariffs to Promote Renewable Electricity in European Countries," ECN-C--02-083, European Competition Network, 2002, https://www.researchgate.net/profile/Jos-Sijm/publication/237535870_The_Performance_of_Feed-in_Tariffs_to_Promote_Renewable_Electricity_in_European_Countries/links/5f034232a6fdcc4ca44eb6b9/The-Performance-of-Feed-in-Tariffs-to-Promote-Renewable-Electricity-in-European-Countries.pdf.

55. J. Hentschel, "Results of the 100,000-roofs solar power programme—an interim review" ("Foerderergebnisse des 100.000 Daecher-Solarstrom-Programms—eine Zwischenbilanz"), ETDWEB (Germany), made available by the US Department of Energy, 2003.

56. Joern Hoppman, Joern Huenteler, and Bastien Girod, "Compulsive policy making: The evolution of the German feed-in tariff system for solar photovoltaic power," *Research Policy* 43 (2014): 1422, 1441.

57. Arne Jungjohann, personal interview, February 15, 2023.

58. Statista, "Investments in renewable energy plants in Germany from 2001 to 2021 (in billion euros)," Statista.com, 2021, https://www-statista-com.stanford.idm.oclc.org/statistics/583526/investments-renewable-energy-plants-germany/.

59. Thure Traber, Claudia Kemfert, and Jochen Diekmann, "German Electricity Prices: Only Modest Increase Due to Renewable Energy Expected," German Institute for Economic Research, Weekly Report 20/11, Berlin, March 16, 2011, https://www.diw.de/sixcms/media.php/73/diw_wr_2011-06.pdf.

60. Statista, "Average retail price of electricity to the residential sector in the United States in selected years from 1975 to 2022(in U.S. cents per kilowatt hour)," Statista.com, December 1, 2023, https://www.statista.com/statistics/183700/us-average-retail-electricity-price-since-1990/.

61. Alex Hongliang et al., "An analysis of the factors driving utility-scale solar PV investments in China: How effective was the feed-in tariff policy?" *Energy Policy* 167 (2022): 113044.

62. Liang-Cheng Ye, João Rodrigues, and Hai Xiang Lin, "Analysis of feed-in tariff policies for solar photovoltaic in China 2011–2016," *Applied Energy* 203 (2017): 496–505.

63. Caolán Magee, "China is set to shatter its wind and solar target five years early, new report finds," CNN, June 29, 2023, https://www.cnn.com/2023/06/29/asia/china-solar-wind-energy-coal-climate-intl/index.html/.

64. Canggui Dong, Runmin Zhou, and Jiaying Li, "Rushing for Subsidies: The impact of feed-in tariffs on solar photovoltaic capacity development in China," *Applied Energy* 281 (2021): 116007.

65. Zvi Lando, CEO, Solar Edge, interview, February 6, 2023.
66. Benjamin Wehrmann, "What German households pay for electricity," *Clean Energy Wire*, January 16, 2023, https://www.cleanenergywire.org/factsheets/what-german -households-pay-electricity.
67. State of California, Office of the Governor, "ICYMI: California Grid Reaches 5,600 MW of Battery Storage Capacity, a 1020% Increase Since 2020," July 12, 2023, https:// www.gov.ca.gov/2023/07/12/icymi-california-grid-reaches-5600-mw-of-battery-stor age-capacity-a-1020-increase-since-2020/.
68. Steve Hanley, "Decoding the Changes to Solar Net Metering Rules In California," *Cleantechnika*, August 19, 2023, https://cleantechnica.com/2023/08/18/decoding-the -changes-to-solar-net-metering-rules-in-california/.
69. Kavya Balaraman, "Sunrun, others adapt as California's new net metering rules spur booming interest in energy storage," *UtilityDive*, April 20, 2023, https://www.utility dive.com/news/sunrun-others-adapt-as-californias-new-net-metering-rules-spur -booming-in/648022/.
70. David Arfin, personal communication, November 8, 2023.
71. Hanley, "Decoding the Changes To Solar Net Metering Rules In California."
72. Tim Hade, interviewed on *Catalyst with Shayle Kann*, October 27, 2023, https://www. canarymedia.com/podcasts/catalyst-with-shayle-kann/the-market-for-microgrids.
73. Jeff Shephard, "Siemens to Divest Solar Business: Focus Renewable Energy Activities on Wind and Hydro Power," *EE Power*, October 22, 2012, https://eepower.com/news/ siemens-to-divest-solar-business-focus-renewable-energy-activities-on-wind-and- hydro-power/.
74. Johan Lilliestam, Helmholtz Centre Potsdam, personal interview, February 24, 2023.
75. Rebecca Brill and Allie Ogletree, "How Much Do Solar Panels Cost in 2023?" *Forbes*, July 25, 2023, https://www.forbes.com/home-improvement/solar/cost-of-solar-panels/.
76. CPI Inflation Calculator, Value of $1 from 2000 to 2023, https://www.in2013dollars. com/us/inflation/2000?amount=1, accessed December 1, 2023.
77. Benjamin Goldstein, Dimitrios Gounaridis, and Joshua Newell, "The carbon footprint of household energy use in the United States," *Proceedings of the National Academy of Sciences* 117, no. 32 (2020): 19122–30.
78. US Environmental Protection Agency, "Heat Pumps," March 6, 2023, https://www. epa.gov/burnwise/heat-pumps.
79. Port PC, "2022 Was the Year of Heat Pumps in Poland," European Heat Pump Associ- ation, 2023, https://www.ehpa.org/port-pc-2022-was-the-year-of-heat-pumps-in-poland/.
80. Clean Energy Council, "Clean Energy Australia Report, 2023," https://assets.clean energycouncil.org.au/documents/Clean-Energy-Australia-Report-2023.pdf.
81. L. Granwal, "Share of energy acquired from renewable sources in Australia in 2022, by state," Statista.com, 2023, https://www.statista.com/statistics/1087804/ australia-renewable-energy-penetration-by-state/.
82. Clean Energy Regulator, "How the scheme works," Australian Government, May 15, 2023, https://www.cleanenergyregulator.gov.au/RET/About-the-Renewable-Energy -Target/How-the-scheme-works.
83. Clean Energy Regulator, "The small-scale technology percentage," Australian Gov- ernment, February 3, 2023, https://www.cleanenergyregulator.gov.au/RET/Scheme -participants-and-industry/the-small-scale-technology-percentage.

84. Trevor Berrill, personal communication, July 3, 2023.
85. Zachary Shahan, "Nearly 1 In 3 Homes in Australia Covered in Solar Panels," *Clean Technica,* February 28, 2023, https://cleantechnica.com/2023/02/28/nearly-1-in-3 -homes-in-australia-covered-in-solar-panels/.
86. Macrotrends, "Australia Carbon (CO_2) Emissions 1990–2023," https://www.macro trends.net/countries/AUS/australia/carbon-co2-emissions.
87. Dan van Holst Pellekaan, "Solar plus storage, the low-emissions solution for SA homes," Ministry for Energy and Mining, South Australia, 2023, https://www.cefc .com.au/where-we-invest/case-studies/solar-plus-storage-the-low-emissions-solution -for-sa-homes/.
88. Ibid.
89. Jeff Sykes, "Is home solar power in Australia still worth it in 2023?" *Solar Choice*, March 14, 2023, https://www.solarchoice.net.au/blog/solar-power-in-australia-is-solar-worth-it/.
90. Jane O'Sullivan, personal communication, June 26, 2023.
91. Ibid.
92. Worldometer, "CO_2 Emissions per Capita," https://www.worldometers.info/co2-emis sions/co2-emissions-per-capita/, accessed July 3, 2023.
93. Heather Kiggin, "What's Next for Carbon Contracts for Difference and Can They Really Boost Innovation in Europe?" European Chemical Industry Council, 2023, https://cefic.org/media-corner/newsroom/whats-next-for-carbon-contracts-for- difference-and-can-they-really-boost-innovation-in-europe/.
94. Bloomberg, "Carbon Contracts for Difference: The Netherlands," 2021, https:// www.bloomberg.com/netzeropathfinders/best-practices/carbon-contracts-for -difference-the-netherlands/.
95. Benjamin Wehrmann, "Carbon Contracts for Difference could kickstart German in- dustry decarbonisation—think tank," *Clean Energy Wire*, 2022, https://www.clean energynon.org/news/carbon-contracts-difference-could-kickstart-german-industry -decarbonisation-think-tank.

Chapter 6. Forcing Climate Technology

1. Union Electric Company v. US Environmental Protection Agency, 427 U.S. 246 (1976).
2. Gary Guzy, "Reconciling Environmentalist and Industry Differences: The New Cor- porate Citizenship 'Race to the Top'?," *Journal of Land Use and Environmental Law* 17 (2002): 409.
3. Arnold Reitze, "Mobile Source Air Pollution Control," *The Environmental Lawyer* 6 (2000): 309.
4. David Gerard and Lester Lave, "Implementing technology-forcing policies: The 1970 Clean Air Act Amendments and the introduction of advanced automotive emission controls in the United States," *Technology Forecasting and Social Change* 72 (2005): 761–78.
5. William Dietrich, "Harmonization of Automobile Standards Under International Trade Agreements: Lessons from the European Union to the WTO and NAFTA," *William and Mary Environmental Law and Policy Review* 20 (1996): 175.
6. T. Shibata, "The influence of big industries on environmental policies: The case of car exhaust standards," in *Environmental Policy in Japan*, ed. S. Tsuru and H. Weidner (Ber- lin: Sigma Bohn, 1989).

7. Tara Bunke, "Environmental Upgrade: The Potential for Chile to Use Market Incentives in Preparing for NAFTA Accession," *Colorado Journal of International Environmental Law and Policy* (1997): 165.

8. Gerard and Lave, "Implementing technology-forcing policies."

9. Gustavo Collantes and Daniel Sperling, "The origin of California's zero-emission vehicle mandate," *Transportation Research Part A: Policy and Practice* 42, no. 10 (2008): 1302–13.

10. Tanvir Arfin et al., "Alternatives to POPs for a healthy life," *Persistent Organic Pollutants*, ed. Kanchan Kumari (Boca Raton, FL: CRC Press, 2021).

11. Y. Zvirin and S. Zamkow, "Solar Energy in Israel: Utilization and Research," Israel–USSR Energy Conference, Technion, Israel Institute of Technology, 1991, http://large.stanford.edu/courses/2016/ph240/kornberg1/docs/25011806.pdf.

12. Greg Whitburn, "History of Solar Energy," *Exploring Green Technology*, 2012, https://exploringgreentechnology.com/solar-energy/history-of-solar-energy/.

13. Rhonda Winter, "Heating water usually accounts for 40 percent of an average family's monthly energy costs," Reuters, March 8, 2011, https://www.reuters.com/article/idUS311612153620110318.

14. Yael Bar-Ilan, David Pearlmutter, and Alon Tal, *Building Green: Policy Mechanisms for Promoting Energy Efficiency in Buildings in Israel* (Haifa: The Technion, Center for Urban and Regional Studies Press, 2010).

15. Madanjeet Singh, *The Timeless Energy of the Sun* (San Francisco: Sierra Club, 1998).

16. Interview with Rami Tarbulaki, CEO of Nimrod Solar and Yisol Chair, January 31, 2023; see: http://www.nimrod-solar.com/en/about/.

17. Wei Li, Tzameret Rubin, and Paul Onyina, "Comparing Solar Water Heater Popularization Policies in China, Israel, and Australia: The Roles of Governments in Adopting Green Innovations," *Sustainable Development* 21, no. 3 (2013): 160–70.

18. G. Zissis, P. Bertoldi, and T. Serrenho, *Update on the Status of LED-Lighting world market since 2018* (Brussels: European Commission, 2021), file:///Users/alontal/Downloads/status_of_led_lighting_world_market_2020_final_rev_2.pdf.

19. US Department of Energy, "Energy Saver," https://www.energy.gov/energysaver/led-lighting, accessed January 26, 2023.

20. P. W. Schultz et al., "Using Social Marketing to Spur Residential Adoption of ENERGY STAR-Certified LED Lighting," *Social Marketing Quarterly* 21, no. 2 (2015), 61–78.

21. Tim Johnston, "Australia Is Seeking Nationwide Shift to Energy-Saving Light Bulbs," *New York Times*, February 21, 2007, https://www.nytimes.com/2007/02/21/business/worldbusiness/21light.html.

22. Nicholas A. A. Howarth and Jan Rosenow, "Banning the bulb: Institutional evolution and the phased ban of incandescent lighting in Germany," *Energy Policy* 67 (2014): 737–46.

23. Ken Lane, *Lighting Sector Report*, International Energy Agency, 2022, https://www.iea.org/reports/lighting.

24. European Commission, "Frequently Asked Questions about the Regulation on Ecodesign Requirements for Non-directional Household Lamps," MEMO/09/113, March 18, 2009.

25. Andrew Stryker and Kathleen Gaffney, "Why the light bulb is no longer a textbook example for price elasticity: Results from choice experiments and demand modeling research," ECCEE, Summer Study, 2015, file:///Users/alontal/Downloads/7-192-13_Stryker.pdf.

26. S. Moghavvemi et al., "Feelings of guilt and pride: Consumer intention to buy LED lights," *PLoS One* 15, no. 6 (June 25, 2020): e0234602, doi: 10.1371/journal.pone.023 4602; PMID: 32584847; PMCID: PMC7316250.

27. "Energy Fact—Cost of LED lighting dropped 15-times as volumes increased," Freeing Energy, https://www.freeingenergy.com/facts/led-bulb-light-cost-price-historical -decline-g213/, accessed January 31, 2023.

28. Research and Markets, "LED Lighting Market Size, Share & Trends: Analysis Report by Product (Lamps, Luminaires), by Application (Indoor, Outdoor), by End Use (Commercial, Residential, Industrial), by Region, and Segment Forecasts, 2022–2030," 2022, https://www.researchandmarkets.com/reports/4538750/led-lighting-market-size -share-and-trends.

29. Jim Vallette, *The New Coal: Plastics and Climate Change* (Bennington, VT: Beyond Plastics, 2021, https://static1.squarespace.com/static/5eda91260bbb7e7a4bf528d8/t/616e f29221985319611a64e0/1634661022294/REPORT_The_New-Coal_Plastics_and_ Climate-Change_10-21-2021.pdf.

30. Paul Rikhter et al., "Life Cycle Environmental Impacts of Plastics: A Review," US Department of Commerce, 2022, https://nvlpubs.nist.gov/nistpubs/gcr/2022/NIST .GCR.22-032.pdf.

31. UNFCCC, "Plastic Promise?," March 22, 2022, https://unfccc.int/blog/plastic-promise.

32. CO_2 Everything, "Plastic Bag (Single Use)," https://www.co2everything.com/, accessed January 27, 2023.

33. Habits of Waste, "Plastic," Habits of Waste website, https://habitsofwaste.org/call-to -action/plastic/, accessed January 27, 2023.

34. Taylor Covington, "Average miles driven per year in the U.S.," *The Zebra*, 2022, https:// www.thezebra.com/resources/driving/average-miles-driven-per-year/.

35. Laura Sullivan, "Recycling plastic is practically impossible—and the problem is getting worse," National Public Radio, October 24, 2022, https://www.npr.org/2022 /10/24/1131131088/recycling-plastic-is-practically-impossible-and-the-problem-is -getting-worse.

36. Christine Koetz, "Plastic bags are not so easily recycled," *Baltimore Sun*, September 26, 2019, https://www.baltimoresun.com/opinion/readers-respond/bs-ed-rr-recycle-plastic -letter-20190926-mp5qh5cayzghnes73w5f2ok7nq-story.html.

37. Bill Gates, *How to Avoid a Climate Disaster* (New York: Knopf, 2021).

38. Eyder Peralta, "Using Plastic Bags Is Now Illegal—and Punishable by Jail Time— in Kenya," National Public Radio, August 28, 2017, https://www.npr.org/sections /thetwo-way/2017/08/28/546680679/using-plastic-bags-is-now-illegal-and -punishable-by-jail-time-in-kenya.

39. Katharine Houreld, "Kenya imposes world's toughest law against plastic bags," Reuters, August 28, 2017, https://www.reuters.com/article/uk-kenya-plastic-idUKKCN1B80PZ.

40. Davina Ngei, "Reflecting on Kenya's single-use plastic bag ban three years on," World Economic Forum, November 25, 2020, https://www.weforum.org/agenda /2020/11/q-a-reflecting-on-kenyas-single-use-plastic-bag-ban-three-years-on/.

41. Aqil Haziq Mahmud, "In Focus: Experiencing Rwanda's plastic bag ban, and whether Singapore could adopt a similar approach," Channel News Asia, July 23, 2022, https:// www.channelnewsasia.com/world/plastic-bag-ban-charge-waste-rwanda-2767096.

42. L. Di Paolo et al., "Carbon Footprint of Single-Use Plastic Items and Their Substitution," *Sustainability* 14 (2022): 16563.

43. UK Department of Environment, Food and Rural Affairs, "Far-reaching ban on single-use plastics in England," 2023, https://www.gov.uk/government/news/far-reaching-ban-on-single-use-plastics-in-england.

44. Renee Cho, "The Truth about Bioplastics," *State of the Planet*, Columbia Planet School, 2017, https://news.climate.columbia.edu/2017/12/13/the-truth-about-bioplastics/.

45. Mohammed Sabbah and Raffaele Porta, "Plastic pollution and the challenge of bioplastics," *Journal of Applied Biotechnology & Bioengineering* 2, no. 3(2017): 111.

46. J. Popp et al., "The effect of bioenergy expansion: Food, energy, and environment," *Renewable and Sustainable Energy Reviews* 32 (2014): 559–78.

47. Environmental Defense Fund, "Avoiding Conversion of Rangelands, 2019, https://www.edf.org/ecosystems/avoiding-conversion-rangelands; see also: Rattan Lal, "Soil carbon dynamics in cropland and rangeland," *Environmental Pollution* (2002): 353–62.

48. S. Walker and R. Rothman, "Life cycle assessment of bio-based and fossil-based plastic: A review," *Journal of Cleaner Production* 261 (2020): 121158.

49. Ghada Atiwesh et al., "Environmental impact of bioplastic use: A review," *Heliyon* 7, no. 9 (2021).

50. Daniel Posen et al., "Greenhouse gas mitigation for U.S. plastics production: Energy first, feedstocks later," *Environmental Research Letters* 12, no. 3 (2017): 034024.

51. Research and Markets, "Surging Online Food Delivery Sector Boosting Demand for Biodegradable Plastics," Global News Wire, October 20, 2022, https://www.globenewswire.com/en/news-release/2022/10/20/2538081/28124/en/Biodegradable-Plastic-Global-Market-Report-2022-Surging-Online-Food-Delivery-Sector-Boosting-Demand-for-Biodegradable-Plastics.html.

52. Aanindeeta Banerjee, personal interview, February 14, 2023.

53. Amy Frankhouser, personal interview, February 14, 2023.

54. Gerard and Lave, "Implementing technology-forcing policies."

55. Anna Bergek and Christian Berggren, "The Impact of Environmental Policy Instruments Innovation: A Review of Energy and Automotive Industry Studies," *Ecological Economics* 10, no. 6 (2014): 112–23.

56. Gerard and Lave, "Implementing technology-forcing policies."

57. J. H. Wesseling, J. C. M. Farla, and M. P. Hekkert, "Exploring car manufacturers' responses to technology-forcing regulation: The case of California's ZEV mandate," *Environmental Innovation and Societal Transitions* 16 (2015): 87–105.

58. Rob Schmitz, "Germany's Green Party wants people to use heat pumps to save energy. Some are balking," National Public Radio, June 21, 2023, https://www.npr.org/2023/06/21/1182636622/germany-heat-pumps-climate-energy.

59. Nikolaus Kurmayer, "Germany adopts watered-down fossil boiler ban for 2028," *Euractiv*, September 11, 2023, https://www.euractiv.com/section/energy-environment/news/germany-adopts-watered-down-fossil-boiler-ban-for-2028/.

60. Gerard and Lave, "Implementing technology-forcing policies."

61. Andres Nentjes, Frans Vries, and Doede Wiersma, "Technology-Forcing through Environmental Regulation," *European Journal of Political Economy* 23 (2007): 903–16.

62. Åsa Grytli et al., "Solar feed-in tariffs and the merit order effect: A study of the German electricity market," *Energy Policy* 61 (2013): 761–70.

63. Baden-Württemberg Climate Protection Act, October 21, 2021, https://www.landesrecht-bw.de/jportal/portal/t/fsa/page/bsbawueprod.psml/action/portlets.jw.MainAction.

64. Anna Ivanova, "Solar roofs to become mandatory for new residential buildings in Baden-Wuerttemberg," *Renewables Now*, April 1, 2022, https://renewablesnow.com/news/solar-roofs-to-become-mandatory-for-new-residential-buildings-in-baden-wuerttemberg-779392, accessed February, 2023/.

65. Peter Park, "Solar obligation for parking spaces: What applies in which federal state?," Peter Park website, May 25, 2022, https://en.peter-park.de/blog/solar-obligation-for-parking-lots-what-applies-in-which-state.

66. California Energy Commission, "Solar PV Systems and Solar Ready," 2023, https://www.energy.ca.gov/programs-and-topics/programs/building-energy-efficiency-standards/online-resource-center/solar.

67. Kantaro Komiya, "Tokyo makes solar panels mandatory for new homes built after 2025," Reuters, December 15, 2022, https://www.reuters.com/world/asia-pacific/tokyo-makes-solar-panels-mandatory-new-homes-built-after-2025-2022-12-15/.

68. Alice Tidey, "EU to phase out new combustion engine cars by 2035," *Euronews*, October 10, 2022, https://www.euronews.com/my-europe/2022/10/28/eu-to-phase-out-new-combustion-engine-cars-by-2035.

69. Gilles Lepesant, "Higher Renewable Energy Targets in Germany: How Will the Industry Benefit?," *Briefings de l'Ifri*, January 6, 2023, https://www.ifri.org/en/publications/briefings-de-lifri/higher-renewable-energy-targets-germany-how-will-industry-benefit.

70. Kerstine Appunn, Freja Eriksen, and Julian Wettengel, "Germany's greenhouse gas emissions and energy transition targets," *Clean Energy Wire,* December 21, 2022, https://www.cleanenergywire.org/factsheets/germanys-greenhouse-gas-emissions-and-climate-targets.

71. Antonia Zimmermann and Joshua Posaner, "Germany Backs Phasing Out Combustion Engine Cars by 2035," *Politico*, March 16, 2022, https://www.politico.eu/article/germany-backs-phasing-out-combustion-engine-cars-by-2035.

72. "California will require half of all heavy trucks sold in the state to be electric by 2035, in an aggressive push to clean up the worst polluters," *New York Times*, March 31, 2023.

73. State of California, Executive Order N-79-20, 2020, https://www.gov.ca.gov/wp-content/uploads/2020/09/9.23.20-EO-N-79-20-Climate.pdf?emrc=9f8f26.

74. Perry Richardson, "All UK taxis must be electric by 2024 to work on Free Now app," *Taxi Point*, January 14, 2021, https://www.taxi-point.co.uk/post/all-uk-taxis-must-be-electric-by-2024-to-work-on-free-now-app.

75. Israel Ministry of Environment, "Ministry Publishes Roadmap for Transition to Public Buses with Zero Emissions by 2025," 2021, https://www.gov.il/en/departments/news/city_buses_without_air_pollution_roadmap.

76. Global EV Outlook, "Entering the Decade of the Electric Drive?," 2020.

77. Michael Porter, "America's Green Strategy," *Scientific American* 264 (1991): 168.

78. Zhiyuan Fan and Julio Friedman, "Low-carbon production of iron and steel: Technology options, economic assessment, and policy," *Joule* 5, no. 4 (2021): 829–62.

79. Hal Harvey and Justin Gillis, *The Big Fix: 7 Practical Steps to Save our Planet* (New York: Simon & Schuster, 2022), 75.

80. Keli Bracmort, "The Renewable Fuel Standard (RFS): Cellulosic Biofuels," Congressional Research Service, 2015, https://crsreports.congress.gov/product/pdf/R/R41106/29.

81. David Gerard, personal interview, July 17, 2023.

82. Barry Commoner, "A Reporter at Large: The Environment," *New Yorker,* June 15, 1987, 56.

Chapter 7. The Power of Public Procurement

1. Rob Stewart, "New healthy food option put to the test," California Department of Corrections and Rehabilitation, March 25, 2023, https://www.cdcr.ca.gov/insidecdcr/2023/03/25/new-healthy-food-option-put-to-the-test/Flavor.
2. State of California, 2017, Buy Clean California Act, *Chapter 816, Sec. 3. (A B 262) Effective January 1, 2018.*
3. State of California, 2003, Assembly Bill 498, Public Contract Code 12400-12404, Chap. 6, *Environmentally Preferable Purchasing Act.*
4. Charleen Fain-Keslar, personal interview, July 31, 2023.
5. Department of General Services, 2023, "Buy Clean California Act (BCCA) Requirements," https://www.dgs.ca.gov/RESD/Resources/Page-Content/Real-Estate-Services-Division-Resources-List-Folder/Buy-Clean-California-Act-BCCA-Requirements.
6. Scot Fong, personal interview, July 31, 2023.
7. USA Spending, "Spending Explorer," 2023, https://www.usaspending.gov/explorer/budget_function.
8. CEIC Data, "China Public Consumption: % or GDP," 2023, https://www.ceicdata.com/en/indicator/china/public-consumption--nominal-gdp.
9. Luke Georghiou et al., "Policy instruments for public procurement of innovation: Choice, design and assessment," *Technological Forecasting and Social Change* 86 (2014): 1–12.
10. Kleoniki Pouikli, "Towards Mandatory Green Public Procurement (GPP): Requirements under the EU Green Deal: Reconsidering the role of public procurement as an environmental policy tool," *ERA Forum* 21 (2021): 699–721.
11. European Commission, "What Is Green Public Procurement?," 2023, https://ec.europa.eu/environment/gpp/what_en.htm.
12. Christine Harland, Jan Telgen, and G. Callender, "International research study of public procurement," in *Public Procurement: International Cases and Commentary*, ed. L. Knight et al. (London: Routledge, 2012), 351–51, ISBN: 9780203815250.
13. Richard Baris, *The Role of Public Procurement in Low-carbon Innovation* (Paris: OECD, 2016), https://www.oecd.org/sdroundtable/papersandpublications/The percent20Role percent20of percent20Public percent20Procurement percent20in percent20Low-carbon percent20Innovation.pdf.
14. Birgit Aschhoff and Wolfgang Sofka, "Innovation on demand: Can public procurement drive market success of innovations?" *Research Policy* 38, no. 8 (2009): 1235–47.
15. European Environment Agency, "Building renovation: Where circular economy and climate meet," EEA website, 2022, https://www.eea. europa.eu/publications/building-renovation-where-circular-economy.
16. Mission Possible Partnership, "Low-carbon concrete and construction: A review of green public procurement programmes," World Economic Forum, 2022, https://missionpossiblepartnership.org/wpcontent/uploads/2022/06/LowCarbonConcreteandConstruction.pdf.
17. Aschhoff and Sofka, "Innovation on demand."
18. Andrea Renda et. al., "The Uptake of Green Public Procurement in the Eu27," Submitted to the European Commission, Environment: Centre for European Policy Studies, 2012, https://www.ajsosteniblebcn.cat/the-uptake-of-green-public-procurement-in-the-eu27_29492.pdf.

19. European Commission, "Directive 2014/24/EU of the European Parliament and of the Council of 26 February 2014 on public procurement and repealing Directive 2004/18/EC," https://eur-lex.europa.eu/legal-content/EN/TXT/?uri=celex%3A32014L0024.
20. Pouikli, "Towards Mandatory Green Public Procurement (GPP)."
21. European Commission, "Directive 2014/24/EU."
22. European Commission, "Green Public Procurement," *Green Business* website, https://green-business.ec.europa.eu/green-public-procurement_en, accessed December 4, 2023.
23. The Netherlands, *Tendering Law 2012*, https://wetten.overheid.nl/BWBR0032203/2022-03-02/0.
24. Mariska Verseveld, personal communication, July 22, 2023.
25. "About PIANOo," https://www.pianoo.nl/en/about-pianoo-0, accessed April 18, 2023.
26. Mariska Verseveld, personal communication, July 22, 2023.
27. Ibid.
28. Astrid Nilsson et al, "Green Public Procurement: A key to decarbonizing construction and road transport in the EU, Stockholm Environmental Institute," 2023, https://www.sei.org/wp-content/uploads/2023/02/green-public-procurement-eu.pdf.
29. Global Lead City Network on Sustainable Procurement, "Rotterdam—the Netherlands," 2023, https://glcn-on-sp.org/cities/rotterdam/.
30. Ibid.
31. Lithuania Public Procurement Office, "Light Board," https://vpt.lrv.lt/lt/svieslente, accessed April 20, 2023.
32. Karolis Kinčius, personal communication, August 4, 2023.
33. National Audit Office of Lithuania, "Are we prepared to conduct green procurement 100%?" (press release), September 8, 2022, https://www.valstybeskontrole.lt/EN/Post/17211/the-national-audit-office-assessed-the-readiness-for-solely-green-public-procurement-from-2023-onwards.
34. Lithuania Environment Ministry, "Report on 2022 Green Public Procurement Share and Activities of the Ministry of Environment of the Republic of Lithuania," May, 2023, https://am.lrv.lt/uploads/am/documents/files/2022%20met%C5%B3%20%C5%BEali%C5%B3j%C5%B3%20vie%C5%A1%C5%B3j%C5%B3%20pirkim%C5%B3%20pa%C5%BEangos%20ataskaita.pdf.
35. Karolis Kinčius, personal communication, August 4, 2023.
36. Mark A. Dutz and Yevgeny Kuznetsov, eds., *Making Innovation Policy Work: Learning from Experimentation* (Washington, DC: OECD/World Bank, 2014), 193–229, https://read.oecd-ilibrary.org/science-and-technology/making-innovation-policy-work_9789264185739-en.
37. Sofia Lundberg and Per-Olov Marklund, "Green public procurement as an environmental policy instrument: Cost-effectiveness," *Environmental Economics* 4, no. 4 (2013).
38. Nilsson et al., "Green Public Procurement."
39. "About Us," Open Contracting Partnership website, https://www.open-contracting.org/about/, accessed August 21, 2023.
40. Open Contracting Partnership, "Open, Sustainable Government Procurement for People, Planet, and Prosperity," policy brief, 2022, https://www.open-contracting.org/wp-content/uploads/2023/02/OCP2023-Policy-brief_-Open-Sustainable-Government-Procurement.-For-People-Planet-and-Prosperity.pdf.

41. "Notice of New Requirements for Concrete," City of Portland, 2019, https://www.port
     land.gov/sites/default/files/2020/concrete-epd-requirements-final-20190515.pdf.
42. "Green Public Procurement: Catalysing the Net-Zero Economy," Mission Possible
     Partnership, World Economic Forum, 2022, https://www3.weforum.org/docs/WEF_
     Green_Public_Procurement_2022.pdf.
43. "Targets," Global Lead City Network on Sustainable Procurement, 2023, https://glcn
     -on-sp.org/cities/helsinki/.
44. Global Lead City Network, "Rotterdam—the Netherlands."
45. Open Contracting Partnership, "Open, Sustainable Government Procurement."
46. Mark Dutz and Dirk Pilat, "Fostering Innovation for Green Growth, Learning Policy
     Experimentation," in Organisation for Economic Co-Operation and Development,
     Making Innovation Policy Work: Learning from Experimentation (Washington, DC:
     OECD/World Bank, 2014), 193–229.
47. European Commission, "Directive 2014/24/EU."
48. European Commission, Live webinar introducing the Strategic Procurement Dia-
     logues project, May 17, 2023, https://public-buyers-community.ec.europa.eu/news/
     live-webinar-introduces-strategic-procurement-dialogues-project.
49. Kostas Selviaridis, Alan Hughes, and Martin Spring, "Facilitating public procurement
     of innovation in the UK defence and health sectors: Innovation intermediaries as insti-
     tutional entrepreneurs," *Research Policy* 52, no. 2 (2023): 04673.
50. Luke Georghiou et al., "Policy instruments for public procurement of innovation:
     Choice, design and assessment," *Technological Forecasting and Social Change* 86 (2014):
     1–12.
51. Ibid.
52. Matti Pihlajamaa and Maria Merisalo, "Organizing innovation contests for public pro-
     curement of innovation: A case study of smart city hackathons in Tampere, Finland,"
     European Planning Studies 29, no. 10 (2021): 1906–24.
53. Cristina Peñasco, Laura Anadón, and Elena Verdolini, "Systematic review of the out-
     comes and trade-offs of ten types of decarbonization policy instruments," *Nature Cli-
     mate Change* 11 (2021): 260.
54. Jean Chemnick, "Kerry pitches climate finance plan. Other countries say it's 'not
     enough,'" *Politico*, November 9, 2022, https://www.politico.com/news/2022/11/09/
     john-kerry-offset-plan-climate-finance-00065753.
55. Varun Sivaram, *Taming the Sun* (Cambridge, MA: MIT Press, 2018).
56. Varun Sivaram, interview in *Catalyst*, podcast with Shayle Kann, December 15, 2022,
     https://www.canarymedia.com/podcasts/catalyst-with-shayle-kann/advance-market
     -commitments-to-decarbonize-heavy-industry.
57. Breakthrough Agenda, *The Breakthrough Agenda Report*, 2022, https://iea.blob.core
     .windows.net/assets/49ae4839-90a9-4d88-92bc-371e2b24546a/THEBREAK
     THROUGHAGENDAREPORT2022.pdf.
58. Varun Sivaram, interview in *Catalyst*.
59. Ibid.
60. Ibid.
61. Mariska Verseveld, Attorney at PIANOo, The Netherlands, personal interview, May
     7, 2023.
62. Ibid.

63. Aure Adell and Bettina Schaefer, *Green Public Procurement in the Republic of Korea: A Decade Of Progress and Lessons Learned,* UNEP, Nairobi, Kenya, 2019, https://wedocs.un ep.org/bitstream/handle/20.500.11822/32535/GPPK.pdf?sequence=1&isAllowed=y.

64. Federal Register, "Federal Acquisition Regulation: Disclosure of Greenhouse Gas Emissions and Climate-Related Financial Risk," November 14, 2022, https://www. federalregister.gov/documents/2022/11/14/2022-24569/federal-acquisition-regula tion-disclosure-of-greenhouse-gas-emissions-and-climate-related-financial.

65. Maxine Joselow, "Biden pushes to require big federal contractors to cut climate pollution," Washington Post, November 10, 2023, https://www.washingtonpost.com/ climate-solutions/2022/11/10/biden-climate-federal-suppliers-cop27/.

Chapter 8. Nudging Down Carbon

1. Felicity Barringer, "Awards Season for Environmentalists," *New York Times*, April 11, 2011, https://archive.nytimes.com/green.blogs.nytimes.com/2011/04/11/awards-season -for-environmentalists/.

2. "Ursula Sladek," The Goldman Prize, 2011, https://www.goldmanprize.org/recipient/ ursula-sladek/, accessed May 3, 2023.

3. Oliver Payne, *Inspiring Sustainable Behaviour: 19 Ways to Ask for Change (London: Routledge, 2012).*

4. Helena Siipi and Polaris Koi, "The Ethics of Climate Nudges: Central Issues for Applying Choice Architecture Interventions to Climate Policy," *European Journal of Risk Regulation* 13 (2022), 218–35.

5. Shui Bin and Hadi Dowlatbadi, "Consumer Lifestyle Approach to Energy Use and the Related CO_2 Emissions," *Energy Policy* 33 (2005): 197–208.

6. M. Crippa et al., "Food systems are responsible for a third of global anthropogenic GHG emissions," *Nature Food* 2 (2021): 198–209.

7. Felix Mormann, "Climate Choice Architecture," *Boston College Law Review* 64 (2023): 1–53.

8. Dan Ariely, *Predictably Irrational: The Hidden Forces That Shape Our Decisions* (New York: HarperCollins, 2010.)

9. Amos Tversky and Daniel Kahneman, "Judgment under Uncertainty: Heuristics and Biases," *Science* 185, no. 4157 (1974): 1124–31.

10. Hilary Byerly et al., "Nudging pro-environmental behavior: Evidence and opportunities," *Frontiers in Ecology* 16, no. 3 (2018): 159–68.

11. Katie Baca-Motes et al., "Commitment and Behavior Change: Evidence from the Field," *Journal of Consumer Research* 39, no, 5 (2013): 1070–84.

12. Noah Goldstein, Robert Cialdini, and Valdas Griskevicius, "A Room with a Viewpoint: Using Social Norms to Motivate Environmental Conservation in Hotels," *Journal for Consumer Research* 35 (2008): 472–82.

13. Daniel Kahneman, *Thinking Fast and Slow* (New York: Penguin, 2012).

14. Daniel Sutter, "The Organ Shortage," American Institute for Economic Research, 2022, https://www.aier.org/article/the-organ-shortage/.

15. Life Source, "Is Organ Donation Rare?" 2021, https://www.life-source.org/latest/is -organ-donation-rare/, accessed May 2, 2023.

16. Richard Thaler and Cass Sunstein, *Nudge: Improving Decisions about Health, Wealth, and Happiness* (New Haven, CT: Yale University Press, 2008), 175–82.

17. Eric Johnson and Daniel Goldstein, "Do Defaults Save Lives?" *Science* (2009): 1338–39.

18. "WHO ONT," Global Observatory on Donation and Transplantation, 2023, https://www.transplant-observatory.org.

19. W. Christian, "Government finally warming up to opt-out organ donations," *Copenhagen Post*, February 1, 2023, https://cphpost.dk/2023-02-01/news/government-finally-warming-up-to-opt-out-organ-donations/.

20. Roberta Fusaro and Julia Sperling-Magron, "Much Anew About Nudges: Interview with Cass Sunstein and Richard Thaler," McKinsey website, August 6, 2021, https://www.mckinsey.com/capabilities/strategy-and-corporate-finance/our-insights/much-anew-about-nudging.

21. Lars Tummers, "Nudge in the news: Ethics, effects, and support of nudges," *Public Administration Review* (2022): 1–22.

22. "Who We Are," Behavioral Insights Team, 2013, https://www.bi.team/, accessed May 7, 2023.

23. "What we do," Competence Centre on Behavioural Insights, European Commission, 2023, https://knowledge4policy.ec.europa.eu/behavioural-insights_en, accessed August 1, 2023.

24. Carsta Simon and Marco Taliabue, "Feeding the behavioral revolution: Contributions of behavior analysis to nudging and vice versa, *Journal of Behavioral Economics for Policy* 2, no. 1 (2018): 91–97.

25. "About SBST," US Social and Behavior Sciences Tea website, accessed May 7, 2023, https://sbst.gov/.

26. Anne Sofie et al., *Nudging and pro-environmental behavior*, Nordic Council of Ministers (Denmark: TemaNord, 2016), https://norden.diva-portal.org/smash/get/diva2:1065958/FULLTEXT01.pdf.

27. T. Kurz, N. Donaghue, and I. Walker, "Utilizing a social-ecological framework to promote water and energy conservation: A field experiment," *Journal of Applied Social Psychology* 35 (2005): 1281–3.

28. Maria Bernedo, Paul Ferraro, and Michael Price, "The Persistent Impacts of Norm-Based Messaging and Their Implications for Water Conservation," *Journal of Consumer Policy* 37 (2014): 437–52.

29. Andrea Waltin and Margee Hume, "Creating Positive Habits in Water Conservation: The Case of the Queensland Water Commission and the Target140 Campaign," *International Journal of Nonprofit and Voluntary Sector Marketing* 16 (2011): 2015–24.

30. Behavioral impacts team, *Impacts of alternatives to In-Home Displays on customers' energy consumption* (London: Open Government License, 2019), https://www.bi.team/wp-content/uploads/2020/10/smart-meters-in-home-displays-impacts-1.pdf.

31. John Lynham et al., "Why does real-time information reduce energy consumption?" *Energy Economics* 54 (2016): 173–81.

32. Mariateresa Silvi and Emilio Rosa, "Reversing impatience: Framing mechanisms to increase the purchase of energy-saving appliances," *Energy Economics* 103 (2021): 105563.

33. Dirk Brounen and Nils Kok, "On the economics of energy labels in the housing market," *Journal of Environmental Economics and Management* 62, no. 2 (2011): 166–79.

34. Sigurd Næss-Schmidt et al., "Do homes with better energy efficiency ratings have higher house prices? Concluding Report," Copenhagen Economics, 2016, https://copenhageneconomics.com/wp-content/uploads/2022/05/copenhagen-economics-2016-do-homes-with-better-energy-efficiency-ratings-have-higher-house-prices.pdf.

35. US Environmental Protection Agency, *Inflation Reduction Act Programs to Fight Climate Change by Reducing Embodied Greenhouse-Gas Emissions of Construction Materials and Products*, 2023, https://www.epa.gov/inflation-reduction-act/inflation-reduction-act -programs-fight-climate-change-reducing-embodied, accessed June 1, 2023.

36. Adrian Camilleri, Dalia Patino-Echeverri, and Rick Larrick, "What's your beef? How 'carbon labels' can steer us towards environmentally friendly food choices," *The Conversation*, December 17, 2018, https://theconversation.com/whats-your-beef-how-carbon -labels-can-steer-us-towards-environmentally-friendly-food-choices-108424.

37. A. R. Camilleri et al., "Consumers underestimate the emissions associated with food but are aided by labels," *Nature Climate Change* 9 (2019): 53–58.

38. K. G. Grunert, S. Hieke, and J. Wills, "Sustainability labels on food products: Consumer motivation, understanding and use," *Food Policy* 44 (2014): 177–89.

39. M. Guenther, C. M. Saunders, and P. R. Tait, "Carbon labeling and consumer attitudes," *Carbon Management* 3 (2012): 445–55.

40. John Lynham et al., "Why does real-time information reduce energy consumption?" *Energy Economics* 54 (2016): 173–81.

41. Felix Mormann, "Climate Choice Architecture," *Boston College Law Review* 64 (2023): 1–53.

42. Robert Cialdini, Raymond Reno, and Carl Kallgreen, "A Focus of Normative Conduct: Recycling the Concept of Norms to Reduce Littering in Public Places," *Journal of Personality and Social Psychology* 58 (1990): 1015–26.

43. United Nations Environmental Program, *UNEP Food Waste Index Report*, 2021, https:// www.unep.org/resources/report/unep-food-waste-index-report-2021.

44. Steffen Kallbekken and Håkon Sælen, "Nudging' hotel guests to reduce food waste as a win–win environmental measure," *Economics Letters* 119, no. 3 (2013): 325–27.

45. Victoria Campbell-Arvai, Joseph Arvai, and Linda Kalof, "Motivating Sustainable Food Choices: The Role of Nudges, Value Orientation, and Information Provision," *Environment and Behavior* 46, no. 4 (2014): 453–75.

46. Daniel Kahneman, *Thinking Fast and Slow* (New York: Penguin, 2012).

47. Cass Sunstein and Lucia Reisch, "Greener by Default," Discussion Paper No. 951, Harvard, John M. Olin Discussion Paper Series, 14, 2018, http://www.law.harvard.edu/ programs/olin_center/papers/pdf/Sunstein_951.pdf.

48. Daniel Pichert and Konstantinos Katsikopoulos, "Green Defaults Information Presentation and Pro-Environmental Behavior," *Journal of Environmental Psychology* 28, no. 1 (2008): 63–73.

49. "Opt Out," San Jose Clean Energy website, https://sanjosecleanenergy.org/opt-out/, accessed May 3, 2023.

50. Felix Ebeling and Sebastian Lotz, "Domestic uptake of green energy promoted by opt-out tariffs," *Nature Climate Change* 5 (2015): 868–71.

51. Cass Sunstein and Lucia Reisch, "Greener by Default," *Trinity College Law Review* 21 (2018): 31–32.

52. Isaac Dinner et al., "Partitioning Default Effects: Why People Choose Not to Choose," *Journal of Experimental Psychology* 17, no. 4 (2011): 332–41.

53. F. Ölander and J. Thøgersen, " Informing Versus Nudging in Environmental Policy," *Journal of Consumer Policy* 37 (2014): 341–56.

54. J. Egebark and M. Ekström, "Can indifference make the world greener?" *Journal of Environmental Economics and Management* 76 (2016): 1–13.
55. Zachary Brown et al., "Testing the effect of defaults on the thermostat settings of OECD employees," *Energy Economics* 39 (2013): 121–34.
56. "Don't Mess with Texas," website, http://www.dontmesswithtexas.org/, accessed May 7, 2023.
57. Ian Ayres, Sophie Raseman, and Alice Shih, "Evidence from Two Large Field Experiments That Peer Comparison Feedback Can Reduce Residential Energy Usage," *Journal of Law, Economics, and Organization* 29, no. 5 (October 2013): 992–1022.
58. Behavioural Insights Team, *Behaviour Change and Energy Use*, UK Cabinet Office, 2011, https://assets.publishing.service.gov.uk/government/uploads/system/uploads/attachment_data/file/60536/behaviour-change-and-energy-use.pdf.
59. Sofie and Nielsen, *Nudging and pro-environmental behavior.*
60. John Stuart Mill, *On Liberty*, Chapter 4, 144, https://www.gutenberg.org/cache/epub/34901/pg34901-images.html.
61. Cass Sunstein, "The Ethics of Nudging," *Yale Journal on Regulation* 32 (2015): 413–15.
62. Sunstein and Reisch, "Greener by Default."
63. Siipi and Koi, "The Ethics of Climate Nudges."
64. David Hagmann, Emily Ho, and George Loewenstein, "Nudging out support for a carbon tax," *Nature Climate Change* 9 (2019): 484–89.
65. US Securities and Exchange Commission, "SEC Proposes Rules to Enhance and Standardize Climate-Related Disclosures for Investors" (press release), 2022, https://www.sec.gov/news/press-release/2022-46.
66. Ibid.
67. Felix Mormann and Milica Mormann, "The Case for Corporate Climate Ratings: Nudging Financial Markets," *Arizona State Law Journal* 53 (2022): 1209–82.

Chapter 9. Disruption

1. Mark Jacobson, curriculum vitae, https://web.stanford.edu/group/efmh/jacobson/vita/, accessed May 15, 2023.
2. Mark Jacobson, Hearing before the Committee on Oversight and Government Reform, House of Representatives, First Session (October 18, 2007), Serial number 110-86, https://www.govinfo.gov/content/pkg/CHRG-110hhrg45164/html/CHRG110hhrg45164.htm.
3. Mark Jacobson and Mark Delucchi, "A Plan to Power 100 Percent of the Planet with Renewables," *Scientific American*, November 1, 2009, https://www-scientificamerican-com.stanford.idm.oclc.org/article/a-path-to-sustainable-energy-by-2030/.
4. Mark Jacobson, *No Miracles Needed: How Today's Technologies Can Save our Climate and Clean Our Air* (Cambridge, UK: Cambridge University Press, 2023).
5. Mark Jacobson, "Abstracts of 89 Peer-Reviewed Published Journal Articles from 25 Independent Research Groups with 142 Different Authors Supporting the Result That Energy for Electricity, Transportation, Building Heating/Cooling, and/or Industry Can Be Supplied Reliably with 100 percent or Near-100 percent Renewable Energy at Different Locations Worldwide," personal website, July 18, 2023, http://web.stanford.edu/group/efmh/jacobson/Articles/I/CombiningRenew/100PercentPaperAbstracts.pdf.

6. National Academy of Science, "PNAS Announces Six 2015 Cozzarelli Prize Recipients" (press release), March 1, 2016, https://www.nasonline.org/news-and-multimedia/news/pnas-cozzarelli-2022.html.

7. Net Zero Tracker, *Net Zero Stocktake 2023*, June, 2023, https://ca1-nzt.edcdn.com/Reports/Net_Zero_Stocktake_2023.pdf?v=1689326892 .

8. Mark Jacobson et al., "100 percent Clean and Renewable Wind, Water, and Sunlight All-Sector Energy Roadmaps for 139 Countries of the World," *Joule* 1, no. 1 (2017), 108–21.

9. Mark Jacobson et al., "Impacts of Green New Deal Energy Plans on Grid Stability, Costs, Jobs, Health, and Climate in 143 Countries," *One Earth* 1 (2019): 449–63.

10. Stella Karagianni, and Maria Pempetzoglou, "The Income Distribution Impact of Decarbonization in Greece: An Initial Approach," *Circular Economy and Sustainability* 2 (2022): 557–67.

11. R. Smits, "Innovation studies in the 21st century: Questions from a user's perspective," *Technological Forecasting and Social Change* 69 (2002): 861–83.

12. Mary Tripsas, "Unraveling the Process of Creative Destruction: Complementary Assets and Incumbent Survival in the Typesetter Industry," *Strategic Management Journal* (1998): 119–42.

13. Greg Daugherty, "The Rise and Fall of Telephone Operators," History Channel website, 2021, https://www.history.com/news/rise-fall-telephone-switchboard-operators.

14. Joseph A. Schumpeter, *Capitalism, Socialism, and Democracy*, 3rd ed. (New York: Harper Perennial, 2008), 83.

15. "About the ILO," International Labor Organization website, https://www.ilo.org/global/about-the-ilo/lang--en/index.htm, accessed May 12, 2023.

16. International Labor Organization, *Working on a warmer planet: The impact of heat stress on labour productivity and decent work* (Geneva: ILO, 2019), https://www.ilo.org/wcmsp5/groups/public/---dgreports/---dcomm/---publ/documents/publication/wcms_711919.pdf.

17. Ibid.

18. Kirk Siegler, "Will skiing survive? Resorts struggle through a winter of climate and housing woes," *Consider This*, National Public Radio, March 31, 2022, https://www.npr.org/2022/03/31/1088236413/will-skiing-survive-resorts-struggle-through-a-winter-of-climate-and-housing-woe.

19. Kimberly Oremus, "Climate variability reduces employment in New England fisheries," *Proceedings of the National Academy of Sciences* 116, no. 52 (2019): 26444–49.

20. Ted Goldammer, "Vineyard Site Selection," in *Grape Grower's Handbook* (Centreville, VA: Apex Publishers, 2018), http://www.wine-grape-growing.com/wine_grape_growing/vineyard_site_selection/vineyard_site_selection_climatic_components.htm.

21. Mike Scott, "Hard-hit by climate change, winemakers turn to sustainability to ride the storms," Reuters, July 29, 2022, https://www.reuters.com/business/sustainable-business/hard-hit-by-climate-change-winemakers-turn-sustainability-ride-storms-2022-09-14/.

22. Aleksander Szpor and Konstancja Ziółkowska, *The Transformation of the Polish Coal Sector*, International Institute for Sustainable Development, 2018, https://www.iisd.org/system/files/publications/transformation-polish-coal-sector.pdf.

23. Krzysztof Kropidlowski, "Poland Nears Coal Glut Prompting Powerful Union to Raise Alarm," *Bloomberg*, August 4, 2023, https://www.bloomberg.com/news/articles/2023

-08-04/poland-nears-coal-glut-prompting-powerful-union-to-raise-alarm?embedded
-checkout=true.

24. US Department of Labor, 2023, *Coal Fatalities for 1900 through 2022*, https://arlweb
.msha.gov/stats/centurystats/coalstats.asp.

25. US Department of Labor, *End Black Lung Now*, USDL website, 2023, https://arlweb
.msha.gov/S&HINFO/BlackLung/homepage2009.asp, accessed May 14, 2023.

26. "Service Stations: FAQs," American Petroleum Institute, 2023, https://www.api.org/
oil-and-natural-gas/consumer-information/consumer-resources/service-station-faqs,
accessed May 14, 2023.

27. David Welch, "When Gas Stations Run Out of Gas," *NBC News*, June 17, 2008, https://
www.nbcnews.com/id/wbna25214948.

28. Zachary Crocket, "Why most gas stations don't make money from selling gas," *The
Hustle*, April 15, 2022, https://thehustle.co/why-most-gas-stations-dont-make-money
-from-selling-gas/.

29. S. R. Moghadam et al., "Effect of occupational exposure to petrol and gasoline components on liver and renal biochemical parameters among gas station attendants: A
review and meta-analysis," *Review of Environmental Health* 35 (2020): 517–30.

30. M. Crippa et al., "Food systems are responsible for a third of global anthropogenic GHG
emissions," *Nature Food* 2 (2021): 198–209.

31. James MacDonald, Jonathan Law, and Roberto Mosheim, *Consolidation in U.S. Dairy
Farming*, US Department of Agriculture, Economic Research Report No. 274, 2020,
https://www.ers.usda.gov/webdocs/publications/98901/err-274.pdf.

32. Nina Lakhani, "US dairy policies drive small farms to 'get big or get out' as monopolies get rich," *The Guardian*, January 31, 2023, https://www.theguardian.com/
environment/2023/jan/31/us-dairy-policies-hurt-small-farms-monopolies-get-rich.

33. "About Remilk," *Remilk's Way* website, https://www.remilk.com/gains, accessed May 14,
2023.

34. Danielle Wiener-Bronner, "Lab-grown meat is cleared for sale in the United States,"
CNN, June 21, 2023, https://www.cnn.com/2023/06/21/business/cultivated-meat-us
-approval/index.html.

35. International Energy Agency, *Renewable Energy and Jobs: Annual Review 2022*, https://
www.irena.org/publications/2022/Sep/Renewable-Energy-and-Jobs-Annual-Re
view-2022.

36. "Rosy for Riveters," *The Economist*, February 4, 2023, 16–19.

37. Organisation for Economic Co-operation and Development, *Employment Implications of Green Growth: Linking jobs, growth, and green policies*, 2017, https://www.oecd
.org/environment/Employment-Implications-of-Green-Growth-OECD-Report-G7-
Environment-Ministers.pdf.

38. International Energy Agency, *World Energy Employment*, September, 2022, https://www
.iea.org/reports/world-energy-employment/executive-summary.

39. "America's Government Is Spending Lavishly to Revive Manufacturing," *The Economist*, February 2, 2023, https://www.economist.com/briefing/2023/02/02/americas-gov
ernment-is-spending-lavishly-to-revive-manufacturing.

40. International Energy Agency, *Electricity Grids and Secure Energy Transitions*, https://
iea.blob.core.windows.net/assets/ea2ff609-8180-4312-8de9-494bcf21696d/Electricity
GridsandSecureEnergyTransitions.pdf.

41. Suzanne Duke, "The future of jobs is green: How climate change is changing labour markets," World Economic Forum, May 1, 2023, https://www.weforum.org/agenda/2023/04/future-of-jobs-is-green-2023-climate-change-labour-markets/.

42. Kate Gordon, Senior Advisor to the Secretary of Energy, Comments at Stanford Graduate School of Business class in public policy and climate innovation, May 8, 2023.

43. Gina Raimondo, "A Moderated Conversation on Semiconductors and the CHIPs and Science Act," US Department of State, 2023, https://www.state.gov/secretary-antony-j-blinken-with-secretary-of-commerce-gina-m-raimondo-indiana-governor-eric-holcomb-senator-todd-young-and-purdue-university-president-mitch-daniels-in-a-moderated-conversatio/.

44. Stephen Graddick, "The US is facing a critical shortage of high-tech engineers," *Scripts*, January 31, 2023, https://scrippsnews.com/stories/us-facing-critical-shortage-of-high-tech-engineers/.

45. Maria Martinez and Riham Alkousaa, "Learn by doing: German renewables companies bid to beat labour shortage," Reuters, April 20, 2023, https://www.reuters.com/business/energy/learn-by-doing-german-renewables-companies-bid-beat-labour-shortage-2023-04-20/.

46. Aaron Atteridge and Claudia Strambo, "Seven principles to realize a just transition to a low-carbon economy," Stockholm Environment Institute, 2020, https://www.sei.org/publications/seven-principles-to-realize-a-just-transition-to-a-low-carbon-economy/.

47. Ibid.

48. Fergus Green and Ajay Gambhir, "Transitional assistance policies for just, equitable, and smooth low-carbon transitions: Who, what and how?" *Climate Policy* 20, no. 8 (2020): 902–21.

49. Ibid.

50. President William Clinton, Executive Order 12898, *Federal Actions to Address Environmental Justice in Minority Populations and Low-Income Populations*, 1994, https://www.archives.gov/files/federal-register/executiveorders/pdf/12898.pdf.

51. John Rawls, *A Theory of Justice* (Cambridge, MA: Harvard University Press, 1972).

52. US Economic Development Administration, *Coal Communities Commitment: Diversifying Coal Communities for a Resilient Future*, 2023, https://www.eda.gov/sites/default/files/2022-10/EDA-Coal-Communities-Commitment-Fact-Sheet.pdf.

53. Senate of the United States, 2022, 26 USC § 48C(c)(1) The Inflation Reduction Act of 2022,117th Congress, 1st Session (H. R. 5376), https://www.democrats.senate.gov/imo/media/doc/inflation_reduction_act_of_2022.pdf.

54. US Department of Energy, "Clean Energy Infrastructure 2023: Qualifying Advanced Energy Project Credit (48C) Program," June 27, 2023, https://www.energy.gov/infrastructure/qualifying-advanced-energy-project-credit-48c-program.

55. Jack Kelly, "Goldman Sachs Predicts 300 Million Jobs Will Be Lost or Degraded by Artificial Intelligence," *Forbes*, March 21, 2023, https://www.forbes.com/sites/jackkelly/2023/03/31/goldman-sachs-predicts-300-million-jobs-will-be-lost-or-degraded-by-artificial-intelligence/.

56. Daron Acemoglu and Pascual Restrepo, "Tasks, Automation, and the Rise in US Wage Inequality," National Bureau of Economic Research, Working Paper no. 28920, 2021.

Chapter 10. Development and Decarbonization

1. "Luc Gnacadja," UN Convention to Combat Desertification, 2023, https://www.unccd.int/convention/who-is-who/unccd-staff/luc-gnacadja, accessed July 7, 2023.

2. "Energy Profile, Benin," International Renewable Energy Agency, August 24, 2022, https://www.irena.org/-/media/Files/IRENA/Agency/Statistics/Statistical_Profiles/Africa/Benin_Africa_RE_SP.pdf.

3. Luc Gnacadja, personal interview, July 6, 2023.

4. Viviane Clement et al., *Groundswell Part 2: Acting on Internal Climate Migration* (Washington, DC: The World Bank, 2021).

5. United Nations Environmental Program, *Evolving Environmental Perceptions: From Stockholm to Nairobi*, ed. Mostafa Kamal Tolba (London: Butterworths, 1988).

6. Alon Tal, *Speaking of Earth: Environmental Speeches That Moved the World* (New Brunswick, NJ: Rutgers University Press, 2006).

7. Alon Tal, "Drinking Coca Cola while the Earth Burns," *Times of Israel*, November 26, 2022, https://blogs.timesofisrael.com/drinking-coca-cola-while-the-earth-burns/.

8. "Climate and Development: An Agenda for Action," World Bank Group, 2022, https://openknowledge.worldbank.org/server/api/core/bitstreams/2df5ceb2-4b98-595f-b384-4b9590c5388a/content.

9. "Implications for African Countries of a Carbon Border Adjustment Mechanism in the EU," African Climate Foundation and the London School of Economics and Political Science, 2023, https://www.lse.ac.uk/africa/assets/Documents/AFC-and-LSE-Report-Implications-for-Africa-of-a-CBAM-in-the-EU.pdf.

10. Jonny Peters, senior policy advisor for trade and *climate*, E3G, personal communication, December 4, 2023.

11. Will de Freitas, "Sahara has huge solar energy potential," World Economic Forum, 2020, https://www.weforum.org/agenda/2020/01/solar-panels-sahara-desert-renewable-energy.

12. Statista, "Solar energy capacity in selected countries in Africa in 2021 (in megawatts)," 2022, https://www.statista.com/statistics/1278125/leading-countries-in-solar-energy-capacity-in-africa/.

13. Benjamin Wehrmann, "Solar power capacity in 2022 sees fast growth, still well below target," *Clean Energy Wire*, February 3, 2023, https://www.cleanenergywire.org/news/german-solar-power-capacity-2022-sees-fast-growth-still-well-below-target.

14. World Bank, "Poverty and Inequality Platform," 2023, pip.worldbank.org, accessed May 29, 2023.

15. United Nations, "Addressing Poverty," UN website, 2023, https://www.un.org/en/academic-impact/addressing-poverty.

16. UN Conference on Trade and Development, *Economic Development in Africa Report, 2021*, 2022, https://unctad.org/press-material/facts-and-figures-7.

17. Charles Palmer Natalie, Pearson, and Georginia Kyriacou, "What is the role of deforestation in climate change and how can 'Reducing Emissions from Deforestation and Degradation' (REDD+) help?" London School of Economics, Grantham Research Institute on Climate Change and the Environment, February 10, 2023, https://www.lse.ac.uk/granthaminstitute/explainers/whats-redd-and-will-it-help-tackle-climate-change/.

18. Global Forest Watch, "Democratic Republic of the Congo," 2023, https://www.globalforestwatch.org/dashboards/country.

19. European Commission, "Emissions Data Base for Global Atmospheric Research, 2022," https://edgar.jrc.ec.europa.eu/report_2022.

20. World Bank, "Access to Electricity (percent of population)—Sub-Saharan Africa, Congo, Dem. Rep.," 2023, https://data.worldbank.org/indicator/EG.ELC.ACCS.ZS?

end=2020&locations=ZG-CD&start=1996&view=chart&year_high_desc=true, accessed May 29, 2023.

21. Sam Lawson, "Illegal Logging in the Democratic Republic of the Congo," Energy, Environment and Resources EER, London, Chatham House, 2014, https://www.chathamhouse.org/sites/default/files/home/chatham/public_html/sites/default/files/20140400LoggingDRCLawson.pdf.

22. World Food Program, "Democratic Republic of the Congo—Emergency," 2023, https://www.wfp.org/emergencies/drc-emergency.

23. Victoria Schneider, "Poor governance fuels 'horrible dynamic' of deforestation in DRC," *Mongabay*, 2020, https://news.mongabay.com/2020/12/poor-governance-fuels-horrible-dynamic-of-deforestation-in-drc/.

24. Wanjira Mathai, personal interview, July 19, 2023.

25. UN Environmental Program, "Why the Global Fight to Tackle Food Waste Has Only Just Begun," 2021, https://www.unep.org/news-and-stories/story/why-global-fight-tackle-food-waste-has-only-just-begun.

26. US Food and Drug Administration, "Food Loss and Waste," 2023, https://www.fda.gov/food/consumers/food-loss-and-waste.

27. Robin Leichenko and Julie Silva, "Climate change and poverty: Vulnerability, impacts, and alleviation strategies," *Wires Climate Change* 5, no. 4 (2014): 539–56.

28. Tariq Khokhar, "Globally, 70 percent of Freshwater Is Used for Agriculture," *World Bank Blogs* 2017, https://blogs.worldbank.org/opendata/chart-globally-70-freshwater-used-agriculture.

29. World Bank, "Agricultural irrigated land (percent of total agricultural land)," 2023, https://databank.worldbank.org/metadataglossary/world-development-indicators/series/AG.LND.IRIG.AG.ZS.

30. Ashok Gulati, Jikun Huang, and Alon Tal, *From Food Scarcity to Surplus: Innovations in Indian, Chinese, and Israeli Agriculture* (New York: Springer Press, 2021).

31. M. Crippa et al., "Food systems are responsible for a third of global anthropogenic GHG emissions," *Nature Food* 2 (2021): 198–209.

32. UN Environment Programme and Climate and Clean Air Coalition, *Global Methane Assessment: Benefits and Costs of Mitigating Methane Emissions* (Nairobi: United Nations Environment Programme, 2021), file:///Users/alontal/Downloads/2021_Global-Methane_Assessment_full_0.pdf.

33. Guillermo Guardia et al., "Subsurface drip irrigation reduces CH4 emissions and ecosystem respiration compared to surface drip irrigation," *Agricultural Water Management* (2023): 285.

34. Landgeist, "Rice Consumption in Africa," November 24, 2022, https://landgeist.com/2022/11/24/rice-consumption-in-africa/.

35. Jane O'Sullivan, "Revisiting Demographic Transition: Correlation and Causation in the Rate of Development and Fertility Decline," in *Proceedings of the 27th International Population Conference*, IUSSP, Busan, Republic of Korea, 2013, 26–31, http://iussp.org/sites/default/files/event_call_for_papers/OSullivan_IUSSP27_Demographic Transition_FullPaper.pdf; see also: Jane O'Sullivan, "Demographic Delusions: World Population Growth Is Exceeding Most Projections and Jeopardising Scenarios for Sustainable Futures," *World* 4 (2023): 545–68.

36. Clarisse Mutimukeye, senior program director, Rwanda Women Doctors for Reproductive Justice, personal interview, July 24, 2023.

37. World Population Review, "DR Congo—Growth," 2023, https://worldpopulation review.com/countries/dr-congo-population, accessed May 30, 2023.

38. Priscilla Achakpa, executive director, Women Environmental Programme, personal interview, June 29, 2023.

39. Gloria Uwingabiye, "Improving access to finance for cash-constrained entrepreneurs in Africa," *Global Dev*, March 21, 2022, https://globaldev.blog/improving-access -finance-cash-constrained-entrepreneurs-africa/.

40. Victor Oluwole, "7 of the worst performing currencies in Africa in 2022," *Business Insider Africa*, November 5, 2022, https://africa.businessinsider.com/local/markets/7 -of-the-worst-performing-currencies-in-africa-in-2022/s5kqzk1.

41. World Bank, "Mobile cellular subscriptions (per 100 people)—Sub-Saharan Africa," 2023, https://data.worldbank.org/indicator/IT.CEL.SETS.P2?locations=ZG, accessed May 30, 2023.

42. Alessandra Salgado, Arina Anisie, and Francisco Boshell, "Pay-As-You-Go Models," Innovation Landscape Brief, International Renewable Energy Association, 2020, https:// www.irena.org/-/media/Files/IRENA/Agency/Publication/2020/Jul/IRENA_Pay-as -you-go_models_2020.pdf.

43. Abhijit Banerjee and Esther Duflo, *Poor Economics: Rethinking Poverty and the Ways to End It* (Washington, DC: Public Affairs, 2012).

44. Ibrahim Mayaki, "Why infrastructure development in Africa matters," United Nations African Renewal, 2019, https://www.un.org/africarenewal/web-features/why-infra structure-development-africa-matters.

45. Vivien Foster and Cecilia Briceno-Garmendia, "Africa's Infrastructure: A Time for Transformation," World Bank, Africa Development Forum, 2010, http://hdl.handle. net/10986/2692.

46. "Rural population, percent—Country rankings," The Global Economy.com, 2023, https://www.theglobaleconomy.com/rankings/rural_population_percent/.

47. David Mutekanga, personal interview, June 12, 2023.

48. International Energy Agency, "Africa in an evolving global context—Key Findings," 2022, https://www.iea.org/reports/africa-energy-outlook-2022/key-findings.

49. Gavin James, "Independent Power Projects Conquer New Territory in Africa," *African Business*, July 8, 2022, https://african.business/2022/07/energy-resources/independ ent-power-projects-conquer-new-territory-in-africa.

50. Kyle Smith et al., *Gang Tackling: Gigawatt Global's Comprehensive Approach to Successful, Sustainable Solar Development in Sub-Saharan Africa*, Stanford Business School, Case Study, 2023, copy with author.

51. Masami Kojima, and Chris Trimble, *Making Power Affordable for Africa and Viable for Its Utilities* (Washington, DC: World Bank, 2016), http://hdl.handle.net/10986/25091.

52. Breakthrough Agenda, *The Breakthrough Agenda Report, 2022*, https://iea.blob.core .windows.net/assets/49ae4839-90a9-4d88-92bc-371e2b24546a/THEBREAK THROUGHAGENDAREPORT2022.pdf.

53. IRENA, "Barefoot College Partnership Leaves Rural Footprint with Skilled Energy Champions," 2022, https://www.irena.org/News/articles/2022/Jun/IRENA-Barefoot -College-partnership-leaves-rural-footprint-with-skilled-energy-champions.

54. Leisa Burrell, "The sky is the limit: Sierra Leone's 'barefoot' women solar engineers," United Nations Industrial Development Organization, 2021, https://www.unido.org/ stories/sky-limit-sierra-leones-barefoot-women-solar-engineers.

55. "Who We Are," Africa Capacity Building Foundation, 2023, https://www.acbf-pact .org/who-we-are/our-offices.

56. Ibid.

57. African Capacity Building Foundation, *Capacity Imperatives for the SDGs: In line with African Union's Agenda 2063* (Harare, Zimbabwe: ACBF, 2019), https://elibrary.acbf pact.org/acbf/collect/acbf/index/assoc/HASHbe86/8916b748/5ddd1aa9/ed.dir/SDG percent20CIM percent20Report percent20web percent20230119.pdf.

58. Transparency International, "Corruption Perceptions Index: Explore the Results," 2023, Transparency.org., https://www.transparency.org/en/cpi/2022, accessed May 19, 2023.

59. Jeffrey Sachs, *The End of Poverty: Economic Possibilities for Our Time* (New York: Penguin, 2005), 236.

60. "About Us," Multilateral Investment Guarantee Agency (MIGA), 2023, https://www .miga.org/about-us.

61. D. Coady et al., "How large are global energy subsidies?" IMF Working Paper 15/105, 2015.

62. "State of Blended Finance 2022, Policy and Research Reports—Climate Edition," *Convergence*, October 26, 2022, https://www.convergence.finance/resource/state-of -blended-finance-2022/view.

63. Sandra Guzman et al., "Climate Finance Needs of African Countries: Climate Policy Initiative," United Nations Disaster Risk Reduction, 2022, https://www.preventionweb .net/news/climate-finance-needs-african-countries.

64. Kannan Lakmeeharan et al., *Solving Africa's Infrastructure Paradox*, McKinsey & Company, March 6, 2020, https://www.mckinsey.com/capabilities/operations/our-insights/ solving-africas-infrastructure-paradox.

65. Renewable Energy Agency and African Development Bank, *Renewable Energy Market Analysis: Africa and Its Regions*, IRENA and AfDB, Abu Dhabi and Abidjan International, 2022, https://www.irena.org/publications/2022/Jan/Renewable-Energy-Mar ket-Analysis-Africa.

66. State of Green, *The Danish Climate Investment Fund*, 2023, https://stateofgreen.com/en/ solution-providers/the-danish-climate-investment-fund/, accessed June 5, 2023.

67. Anton Eberhard et al., "Independent Power Projects in Sub-Saharan Africa: Investment Trends and Policy Lessons," *Energy Policy* 108 (2017): 390–424, https://doi.org /10.1016/j.enpol.2017.05.023.

68. Achref Chibani, *Green Power Politics in North African Countries: Continuity or Change?*, Arab Center, Washington, DC, March 1, 2023, https://arabcenterdc.org/resource/green -power-politics-in-north-african-countries-continuity-or-change/.

69. Ananthakrishnan Prasad et al., "Mobilizing Private Climate Financing in Emerging Market and Developing Economies," *IMF Staff Climate Notes*, International Monetary Fund, 2022, https://www.imf.org/-/media/Files/Publications/Staff-Climate-Notes /2022/English/CLNEA2022007.ashx.

70. Avinash Persaud, *Breaking the Deadlock on Climate: The Bridgetown Initiative*, Groupe d'études Géopolitiques, 2022, https://geopolitique.eu/en/articles/breaking-the -deadlock-on-climate-the-bridgetown-initiative/.

71. International Institute for Sustainable Development, "Africa Energy Guarantee Facility," https://www.iisd.org/credit-enhancement-instruments/institution/africa-energy-guar antee-facility/, accessed June 2, 2023.

72. Maha El Dahan, Kate Abnett, and Sarah Marsha, "Rich nations to meet overdue $100 billion climate pledge this year," Reuters, May 3, 2023, https://www.reuters.com/world /europe/uaes-jaber-urges-donors-deliver-100-bln-pledge-developing-countries-2023 -05-02/.
73. Zia Weise and Zack Colman, "Paris climate finance summit delivers momentum but few results," *Politico*, June 23, 2023, https://www.politico.eu/article/paris-new-global -financing-pact-summit-macron-climate/.
74. Luc Gnacadja, 2023, personal interview, July 6, 2023.
75. Samuel Ramani, "Russia and China in Africa: Prospective Partners or Asymmetric Rivals?," South African Institute for International Affairs, 2019, https://saiia.org.za/ wp-content/uploads/2021/12/Policy-Insights-120-ramani.pdf.
76. "The IMF faces a nightmarish identity crisis, "*The Economist*, April 4, 2023, https://www.economist.com/finance-and-economics/2023/04/04/the-imf-faces-a -nightmarish-identity-crisis.
77. Organisation for Economic Co-operation and Development, *Climate Finance Provided and Mobilised by Developed Countries in 2016–2020: Insights from Disaggregated Analysis, Climate Finance, and the USD 100 Billion Goal* (Paris: OECD Publishing, 2022), https://doi.org/10.1787/286dae5d-en.
78. Sustainable Energy for All, "Coal Power Finance in High-Impact Countries," 2021, https://www.seforall.org/system/files/2022-08/EF-Coal_Finance-SEforALL.pdf.
79. Jeffrey Sachs, *The End of Poverty: Economic Possibilities for Our Time* (New York: Penguin, 2005).
80. African Development Bank, 2023, "Facility for Energy Inclusion Off-Grid Energy Access Fund - P-Z1-FF0-012 - E&S Policy," https://www.afdb.org/en/documents/multi national-facility-energy-inclusion-grid-energy-access-fund-p-z1-ff0-012-es-policy.
81. Debbie Mohnblatt, "Burundi Inaugurates Country's First Utility-Scale Solar Power Field," *Media Line*, 2023, https://themedialine.org/life-lines/burundi-inaugurates -countrys-first-utility-scale-solar-power-field/.
82. Agence Français de Développement, 2023, "Boosting Access to Renewable Energy Across East Africa," https://www.afd.fr/en/actualites/boosting-access-renewable-energy -across-east-africa.
83. World Bank, "The World Bank in Zambia," 2023, https://www.worldbank.org/en/ country/zambia.
84. World Bank, "Unlocking Low-Cost, Large-Scale Solar Power in Zambia," 2023, https:// www.worldbank.org/en/news/feature/2019/05/14/unlocking-low-cost-large-scale -solar-power-in-zambia, accessed May 31, 2023.
85. The White House, "U.S.–Africa Partnership in Supporting Conservation, Climate Adaptation, and a Just Energy Transition," December 13, 2022, https://www.whitehouse. gov/briefing-room/statements-releases/2022/12/13/fact-sheet-u-s-africa-partnership -in-supporting-conservation-climate-adaptation-and-a-just-energy-transition/.
86. Cece Coffey, "Kenya's Clean Energy Transition Gets a Boost from Solar Power," *Insight*, Kleinman Center for Energy Policy, University of Pennsylvania, January 19, 2023, https://kleinmanenergy.upenn.edu/news-insights/kenyas-clean-energy-transition -gets-a-boost-from-solar-power/.
87. Kenya Ministry of Energy, "Kenya Off-Grid Solar Access Project," 2023, https://energy. go.ke/?page_id=7185.

88. Wanjira Mathai, personal interview, July 19, 2023.
89. Harry Clynch, "Green hydrogen at heart of Namibia's Vision 2030," *African Business*, July 10, 2023, https://african.business/2023/07/energy-resources/green-hydrogen-at -heart-of-namibias-vision-2030.
90. Yun Sun, "China's Aid to Africa: Monster or Messiah?" *Brookings East Asia Commentary*, February 7, 2014, https://www.brookings.edu/opinions/chinas-aid-to-africa-monster -or-messiah/.
91. International Energy Agency, *World Energy Outlook 2022: Executive summary*, https:// www.iea.org/reports/world-energy-outlook-2022/executive-summary.
92. David Mutekanga, personal interview, June 12, 2023.

Chapter 11. Reaching Net Zero

1. Rachel Ramirez, "Four alarming charts that show just how extreme the climate is right now," CNN, June 17, 2023, https://www.cnn.com/2023/06/17/world/four-climate -charts-extreme-weather-heat-oceans/index.html.
2. Al Gore, "Al Gore at the Opening of the COP27 World Leaders Summit, UN Climate Change," November 7, 2022, https://www.youtube.com/watch?v=qLTcC7srnLw.
3. Jesse Jenkins, "Implementing the Inflation Reduction Act: Legal & Policy Challenges on the Road to Net Zero," lecture at 2023 GreenTech Conference, Georgetown University, November 8, 2023, https://www.georgetownclimate.org/files/2023-11-08%20 -%20IRA%20and%20the%20Path%20to%20Net-Zero%20-%20Georgetown%20 Law%20Greentech.pdf.
4. UCAR, "Carbon Dioxide Absorbs and Re-emits Infrared Radiation," National Center for Science Education website, https://scied.ucar.edu/learning-zone/how-climate -works/carbon-dioxide-absorbs-and-re-emits-infrared-radiation, accessed December 5, 2023.
5. Jeremiah Budin, "A Chinese Company Is Releasing an Ultra-Cheap EV That Costs Less than Almost Any New Car in America," *The Cool Down*, June 25, 2023, https://www. thecooldown.com/green-tech/byd-seagull-ev-cheap-electric-car/.
6. World Resources Institute, *Creating a Sustainable Food Future—Final Report* (Washington, DC: WRI, 2019), https://files.wri.org/d8/s3fs-public/wrr-food-full-report.pdf.

Alon Tal's career has been a balance between academia, politics, and public interest advocacy. He is presently a visiting professor at the Stanford Graduate School of Business and former chair of the Department of Public Policy at Tel Aviv University. Tal has published hundreds of academic and popular articles and written and/or edited some eleven books on topics involving sustainability. During 2021 and 2022, Tal was a member of the Knesset, Israel's parliament, where he chaired the country's first subcommittee on environment, climate, and health. He has founded several environmental organizations, and at age forty-eight he received a life achievement award from the Israeli government for his contribution to environmental protection.